Land Use Planning, Environmental Protection and Growth Management

The Florida Experience

by
Robert A. Catlin

CRC Press
Taylor & Francis Group
Boca Raton London New York

CRC Press is an imprint of the
Taylor & Francis Group, an **informa** business

First published 1997 by Ann Arbor Press

Published 2019 by CRC Press
Taylor & Francis Group
6000 Broken Sound Parkway NW, Suite 300
Boca Raton, FL 33487-2742

First issued in paperback 2020

ISBN-13: 978-0-367-57942-5 (pbk)
ISBN-13: 978-1-57504-042-4 (hbk)

Visit the Taylor & Francis Web site at
http://www.taylorandfrancis.com

and the CRC Press Web site at
http://www.crcpress.com

Library of Congress Cataloging-in-Publication Data

Land use planning, environmental protection, and growth management: the Florida experience / by Robert A. Catlin

p. cm.

Includes bibliographical references and index.

ISBN 1-57504-042-5

1. Land use--Planning--Environmental aspects--Florida.
2. Regional planning--Environmental aspects--Florida. I. Title.
HD211.F6C38 1997
333.73'13'09759--dc21 g 22724
CIP

Acknowledgments

To Ethel, Janell, and Michelle

About the Author

Dr. Robert A. Catlin is Professor of Urban Planning and Public Policy at the Edward J. Bloustein School of Planning and Public Policy, Rutgers, The State University of New Jersey, New Brunswick. He has been a professional urban planner since 1961, with local governmental positions in Minneapolis, Southern California, Baltimore, Washington, DC, and New York until 1972, when he began an academic career as an assistant professor of urban planning at California State Polytechnic University, Pomona. He has also held faculty and administrative positions at Georgia Institute of Technology, Indiana University, and the State University of Florida before coming to Rutgers in 1992. Dr. Catlin holds a bachelor's degree in City and Regional Planning from Illinois Institute of Technology, a Master of Science in Urban and Regional Planning from Columbia University and a Ph.D. in American Government from the Claremont Graduate School. In 1978 he became a charter member of the American Institute of Certified Planners.

Contents

Part One
Background

Part Two
The Rationale for Case Studies

List of Figures

List of Tables

PART ONE

BACKGROUND

1 Introduction

Ever since the end of World War II, Florida has been forced to deal with the challenge of managing explosive, rapid, and persistent growth. The state was among the last in the nation to become fully settled, and in 1940 there were only 1.9 million residents. By 1950, the population had jumped to 2.8 million permanent residents, and by 1960 there were 4.9 million residents, a figure that grew to 6.8 million in 1970.

By the late 1960s it became clear that Florida's postwar growth was creating serious and possibly irreversible problems. The extensive destruction of wetlands, dune and beach systems, estuaries, the threat to drinking water due to pollution of the Biscayne and Floridan aquifers and other negative consequences caused alarm among a broad spectrum of Florida citizens, including farmers, environmentalists, landowners, influential retirees, and even developers themselves.[1] In the late 1960s a series of crises took place: (1) planning and initiation of construction of a new Miami jetport in an environmentally sensitive area; (2) construction of the Cross-Florida Barge Canal, threatening the wild scenic Oklawaha River; (3) major fires in the Everglades; and (4) a pronounced drought in southeast Florida due mainly to drainage of much of the Everglades for agriculture. These crises helped to create a political environment favorable for protecting the state's fragile ecology.[2] By 1967 the Florida legislature was reapportioned due to the U.S. Supreme Court's Baker v. Carr or "one man, one vote" ruling[3] meaning that urban legislators representing the areas most threatened by uncontrolled growth had new clout in numbers. Finally, in 1970, Reubin Askew, a young, bright and energetic moderate, became Florida's governor. With Askew's leadership, the Florida legislature adopted the Environmental Land and Water Management Acts of 1972.[4] This package of legislation included (1) the Environmental Land and Water Management Act, which provided for identification and preservation of areas of critical concern, such as wetlands, beaches, dunes, water recharge areas, etc., and developments of regional impact criteria, which set standards for major projects generating traffic, pollution, and a need for substantially improved infrastructure; (2) the Water Resources Act, which established five regional water management districts; (3) the Land Conservation Act, which provided for purchase of environmentally endangered lands; and (4) the Florida Comprehensive Planning Act, which would provide a basis for statewide policies guiding long-range social economic and physical growth within the state.

While these acts provided for the identification and preservation of environmentally sensitive land, there was still a need to deal with another related issue, namely, *How to control growth on land suitable for development.* The answer was enactment by the legislature and Governor Askew of the *Local Government Comprehensive Planning Act* in 1975.[5] This act mandated that all cities, towns, and counties first identify a local government agency to produce its comprehensive plan, then prepare and adopt one. Required elements included future land use, traffic, sanitary sewer, solid waste, drainage and water, conservation, recreation and open space, housing, intergovernmental coordination, utilities, and, if applicable, a coastal zone scheme. Ten optional elements provided a wide range of possibilities for local governments to tailor plans to their special needs. These plans were to be reviewed by state government, but there was no provision for state approval.

I arrived in Florida on August 15, 1977 to take a position as associate professor of political science at the University of South Florida. My job was to teach the planning, urban management, and housing courses in the department's new Master of Public Administration program. With bachelor's and master's degrees in planning and a Ph.D. in American government, coupled with 16 years of experience as a professional urban planner, I took an immediate interest in Florida's efforts to manage growth. The Local Government Comprehensive Planning Act of 1975 made Florida one of only two states that demanded a comprehensive plan for all units of local government (Oregon was the other state), and I was eager to see it in operation. Hillsborough County, which had grown from 397,000 persons in 1960 to 490,000 in 1970, and with an estimated 1975 population of 600,000, was experiencing growth at a higher rate than the state as a whole. Its legislative delegation, in an attempt to harness that growth, had a "local bill"[6] passed in 1975 that required completion of a comprehensive plan for the county by December 31, 1977, two years before other local governments in Florida were required to do so. This plan, known as "Horizon 2000," was in the final stages of public review when I came to town. In 1978, I was invited to become a member of the county's Solid Waste Task Force, which was given a mandate to recommend a new landfill site to the County Commission. In 1979, I was asked to join the Hillsborough County Planning Commission. During my three years on this body, I was involved in the process of reviewing several major development proposals with respect to the adopted Horizon 2000 comprehensive plan.

By 1982, the last of the plans mandated by the 1975 Local Government Comprehensive Planning Act (LGCPA) were beginning to be finalized, adopted by their local governments, and reviewed by the State Department of Community Affairs and the various regional planning councils. By mid-1982, 419 of 461 cities and counties had adopted comprehensive plans reviewed by regional and state agencies.[7] By 1984, all units of local government had prepared and adopted plans. However, by this time, scholars and journalists alike had noted many flaws with the LGCPA. These flaws were essentially: (1) there were no means to assure the quality of local plans and land development regulations; (2) there was no mechanism to tie the local governments' plans to implementation regulations; (3) there was virtually no funding on the part of the state to adequately prepare plans and help implement them; (4) there was a failure to develop a state plan, which could have served as a framework for local plans; and (5) there was an extremely weak state "review and comment" process instead of a mandatory consistency review for local plans.[8]

In order to correct these flaws, the 1985 Local Government Comprehensive Planning Land Development Regulation Act (ch. 163, 1985) and the omnibus Growth Management Act were passed by the Florida legislature and signed into law by Governor Robert Graham. Bob Graham was one of the leading pro–growth control legislators in the early 1970s and was primarily responsible for the passage of the 1972 Environmental Land and Water Management Acts and the 1975 LGCPA. This new bill required that all local governments redraft their plans in accordance with the 1984 state plan, which presented detailed goals, objectives, and policies for controlling growth. Instead of a simple "review and comment" function, the State Department of Community Affairs was empowered to review and *approve* all local government plans, and if these plans were found to be out of compliance with state regulations, fines and withholding of state funds could be levied as possible penalties. A provision in the new law known as "R9j5" provided specific criteria for preparing plan elements acceptable to state government. This provision assured that at least some degree of uniform quality would be attained by these new plans. Most importantly, the Growth Management Act of 1985 mandated that no new development could be approved unless adequate public facilities and services (infrastructure) were in place. Known as "concurrency," this feature was the first of its kind in the nation.[9]

Known formally as the 1985 Local Government Comprehensive Planning and Land Development Regulation Act, this legislation took almost two years between its passage and the issuance of regulations enabling local government to prepare plans in accordance with its provisions. By 1987, all local governments were wrestling with the problems of adjusting their comprehensive plans to the requirements of that legislation. Money was a problem. Although the legislature appropriated $10 million in 1986 to help local government offset the cost of preparing new plans,[10] these funds were sharply curtailed when, in a special session of the 1987 legislature, a coalition of conservation Republican and Democratic lawmakers headed by newly elected Governor Bob Martinez repealed several revenue-enhancing measures passed in the previous year, including a tax on services. These cuts caused a $98 million revenue shortfall in 1987–88 and a $144.5 million shortfall in 1988–89.[11] Not only was there limited funding for plan development, but a select state committee estimated in 1987 that due to concurrency requirements $53 billion would be needed by the year 2000 to cover infrastructure needs for new development — $35 billion at the state level and 18 billion at the local level.[12] *Where was the money going to come from to pay for concurrency?* Florida state government had no answers.

Between 1987 and 1992, I assisted a consultant team in the preparation of a neighborhood plan in West Palm Beach initiated as the result of a Development of Regional Impact study for a mammoth redevelopment of a 28-block area just southwest of that city's Central Business District. I have also witnessed firsthand Gainesville and Alachua County's attempt to prepare a comprehensive plan in accordance with the 1985 act, and especially its interaction with the State Department of Community Affairs in gaining that agency's approval. My experience with Florida's attempts to control urban growth as a participant-observer since 1977 leads me to the realization that legislation alone won't be enough to assure managed growth in Florida. I question the accepted explanation for the failure of Florida's LGCPA of 1975. More than once I saw "quality" plans simply ignored by local elected officials in their efforts to appease development interests. While the concurrency requirements of the 1985 Growth Management Act are far-reaching, and certainly a potential means of assuring planned growth, given a lack of funding by federal, state, and local government for infrastructure, are powerful development interests simply going to twiddle their thumbs and keep moneymaking projects on hold? Will the State Department of Community Affairs use the 1985 legislation to hold local governments' feet to the fire, thus making them comply with the legislative intent, especially when in 1986 a so-called "glitch" bill was passed by the legislature that weakened the state's power to find local plans in noncompliance with state and regional guidelines?[13]

This book is an attempt to get underneath the surface issues of growth management problems in Florida, identifying the underlying causes of resistance to planning, and pointing the way to possible alternatives for reaching the desired goal of orderly growth. In doing this, case studies are utilized to explore the legislation's impact not so much in relationship to the plans produced but to how these plans actually impacted public and private sector development proposals.

It seems that by all accounts the 1972 Environmental Land and Water Management Acts have succeeded in preventing worst-case, large-scale environmental degradation. What is yet to be resolved is how best to control sprawl and unplanned growth on the land identified as suitable for various levels and intensities of development. The questions that remain include: (1) What types of spatial land use mix can provide the greatest good for the greatest number? (2) What intensity of development is appropriate for a given site? (3) How can concurrency as called for in the 1985 Growth Management Act be reached given present funding problems and citizen resistance to new taxes?, and most importantly, (4) What mix of public policies and private consciousness-raising is needed in order to meet the first three considerations?

This book opens with two chapters on Florida's history of physical development covering the period of 1850 to 1985, when the Local Government Comprehensive Planning Acts were passed. Following that will be four case studies, all involving implementation of both the 1975 and

1985 acts. The first, entitled *The Hillsborough County Solid Waste Site Selection Controversy of 1978–79: Struggling with a NIMBY,* details the process by which the Solid Waste Task Force that I served as a member of recommended a landfill site and how the County Commission reacted to it, and how that body's decision worked out in the short and long run through the early 1990s. Next is *Old Hyde Park Village: An Example of Transactive Planning.* This case is drawn from my experience as a member of the Hillsborough County Planning Commission, which reviewed this upscale shopping center in the central core of Tampa for compliance with the Horizon 2000 plan and neighborhood plans prepared by local residents, and how after construction the development turned out by 1995. This case is followed by *The West Palm Beach City Center: A Succession of Plans.* Between 1978 and 1988 no less than four different plans were prepared for the greater West Palm Beach Central Business District (CBD), with the schemes being adjusted every time a major new development proposal was announced by the private sector. The focus here is how these successive plans influenced the built environment of West Palm Beach's CBD by the mid 1990s. The last case is entitled *Alachua County: Running To or From Managed Growth?* This case details attempts by Alachua County's (Gainesville) government to prepare a plan that could be approved by the Florida State Department of Community Affairs while at the same time approving a proposed rezoning from agriculture to industry for 100 acres of open land in this county of 180,000 people (1990 census). In all four cases, I combine my observations and experience as a professional urban planner with the relevant literature in that discipline's history, theory, and practice.

The book concludes first by tying together the history of Florida's development and its unique political culture with the four cases. After a review of the 1993 Planning and Growth Management Act, which substantially weakened the 1985 legislation, recommendations for fundamental state policy and private sector changes will be made.

One could ask, "Why analyze the effectiveness of the Florida growth management acts by the use of the case study method? Why not use a quantitative approach such as numbers of plan amendments and zone changes proposed, the percentage approved with or without revisions by the planning staff, and the percentage of planning staff recommendations approved by city and county commissioners?" Certainly this approach has merit, and I am certain that many who have and will write on Florida's

1975 and 1985 growth management acts will use this method.[14] However, I feel that a descriptive approach must be accompanied by a series of detailed case studies in order to portray a clearer picture of what is really taking place.

This book also attempts to take a hard look at planning ethics in action. In Florida, the historical bias against planned growth runs so deep on the part of elected and appointed officials that I have observed on many, many occasions, professional planners tailoring their recommendations to meet the agendas of planning boards and city or county commissions. These bodies are almost always dominated by development interests: builders, realtors, bankers, attorneys, architects, civil engineers, and "wanna-be's," i.e., those individuals who aspire to the status and power of the development community. In the early days of the planning profession — up to the early 1970s — most planners would forthrightly make their best professional recommendations to decision-makers and, if they suffered a series of reversals by that group, they would simply move on to another job where officials might be more amenable. However, over the past 20 years, the proliferation of graduate level programs in planning that eventually crowded the field,[15] a series of major recessions occurring at the same time as changing federal priorities,[16] and in the mid-1990s an antigovernment mood nationally have all served to shrink funding available for planning. As a result, public planners are now much more reluctant to be "professional," because doing so could mean incurring the wrath of individuals who in effect sign their paychecks.[17] Given the move to reduced governmental involvement in planning regulation as per the 1994 conservative revolution both in Florida and nationally, this phenomenon warrants increased observation.[18]

With some planners first anticipating the preferences of a majority of planning board and elected official bodies and then preparing their recommendations accordingly, quantitative analysis may not present an accurate portrayal as to how well growth management is working in Florida. While the vast majority of planners will, I am sure, present proposals to boards and commissions that they feel are professionally correct, I choose to use the case study approach in order to add to the body of knowledge concerning growth management in Florida.

This book is written not only for academics, practitioners, and students involved in urban and regional planning, but more so for the wider group of citizens in Florida, our nation, and elsewhere who are interested in the dynamics of managing urban growth. Florida will struggle with this problem well into the next century. In 1980, Florida's population grew to 9.8 million, a 44% increase over that in 1970. By 1990 Florida had 13 million residents, making it the nation's fourth largest state behind California, New York, and Texas. Estimates show that some time between 2020 and 2030, there will be 22 million Floridians. In addition to the permanent residents, there will be millions of tourists and winter visitors. People have come to Florida since World War II not only for economic opportunity, but to enjoy its special quality of life as well. One must act now in order to protect and preserve that quality of life for future generations of Floridians.

Notes

1. For a review of issues leading to concern for the environment among Florida's citizens, see Luther Carter, *The Florida Experience: Land and Water Policy in a Growth State,* Johns Hopkins University Press (Baltimore), 1974, pp. 117–139; Robert G. Healy, *Land Use and the States,* Johns Hopkins University Press, 1976, pp. 103–138; and John DeGrove, *Land Growth and Politics,* American Planning Association Press, 1984, pp. 99–176.
2. Carter, op. cit. Note 1, pp. 117–137, 265–312.
3. Baker v. Carr, 1962.
4. Prior to the adoption of these acts, a statewide conference on the 1970–71 water crisis called by Governor Askew prepared a report recommending adoption of the American Law Institute's Model Land Development Code: Article 7. See Robert G. Healy, *Land Use and the States,* pp. 109–112.
5. 1975 Florida Laws 257, Florida Statutes 163, 31613243.
6. Florida law allows the state legislature to pass bills that apply only to one city or county. Usually this is done at the request of the "local" legislative delegation. In 1975, the Hillsborough County legislative delegation petitioned the legislature to pass a requirement for Hillsborough County to prepare its plan by December 31, 1977. Other Florida local governments had until December 31, 1979 to do so.
7. DeGrove, op. cit. Note 1, p. 162.
8. See John M. DeGrove and Nancy E. Stroud, "New Development and Future Trends in Local Government Comprehensive Planning," *Stetson Law Review,* Vol. XVII No. 3, Summer 1989, pp. 573–605; John DeGrove, "Balanced Growth in Florida: A Challenge for Local, Regional and State Governments" in *New Jersey Bell Journal,* Volume 10, No. 3, 1987, pp. 38–44.

Table 1. Florida Population: Census Counts and Urban-Rural Breakdown, 1830–1990.

Year	Total No.	Population Change from # Preceding Census	Percent Change	Urban #	Urban %	Rural #	Rural %
1830	34,730	X	X	0	0.0	34,730	100
1840	54,477	19,747	56.9	0	0.0	54,477	100
1850	87,445	32,968	60.5	0	0.0	87,445	100
1860	140,424	52,979	60.6	5,708	4.1	134,716	95.9
1870	187,748	47,324	33.7	15,275	8.1	172,473	91.9
1880	269,493	81,745	43.5	26,947	10.1	242,546	90.0
1890	391,422	121,929	45.2	77,358	19.8	314,064	80.2
1900	528,542	137,120	35.0	107,031	20.3	421,511	79.7
1910	752,619	224,077	42.4	219,080	29.1	533,539	70.9
1920	968,470	215,851	28.7	353,515	36.5	614,955	63.5
1930	1,468,211	499,741	51.6	759,778	51.7	708,433	48.3
1940	1,897,414	429,203	29.2	1,045,791	55.1	851,623	44.9
1950	2,771,305	873,891	46.1	1,566,788	56.5	1,204,517	43.5
1960	4,951,560	2,180,255	78.1	3,077,989	62.2	1,873,571	37.8
1970	6,791,418	1,839,858	37.2	5,544,551	81.7	1,244,892	18.3
1980	9,746,324	2,954,906	43.5	8,212,385	84.3	1,533,393	15.7
1990	12,937,926	3,191,602	32.7	11,828,476	91.4	1,109,450	8.6

Table 2. Population Growth in Florida's 25 Largest "Urban Counties," 1960–1990[19] (Rank by 1990 Population).

County	Largest City	1960	1970	1980	1990	% Change 1960–1990
Dade	Miami	935,047	1,267,792	1,625,979	1,937,094	107.2
Broward	Ft. Lauderdale	333,964	620,100	1,014,043	1,255,488	275.9
Palm Beach	W. Palm Beach	228,106	348,993	573,125	863,518	278.7
Pinellas	St. Petersburg	374,665	522,329	728,406	851,659	127.5
Hillsborough	Tampa	397,788	490,265	646,960	834,054	109.6
Orange	Orlando	263,540	344,311	471,660	677,491	156.8
Duval	Jacksonville	455,411	528,865	570,981	672,971	47.8
Polk	Lakeland	195,139	228,026	321,652	405,382	107.8
Brevard	Melbourne	111,435	230,006	272,959	398,978	259.0
Volusia	Daytona Beach	125,319	169,487	258,762	370,712	196.3
Lee	Ft. Myers	54,539	105,216	205,266	335,113	515.3
Seminole	Sanford	54,947	83,692	179,752	287,529	422.5
Pasco	Newport Richey	36,785	75,995	194,193	281,131	666.3
Sarasota	Sarasota	76,895	120,413	202,251	277,776	261.1
Escambia	Pensacola	173,829	205,334	233,794	262,798	51.1
Manatee	Bradenton	69,169	97,115	148,442	211,707	206.1
Marion	Ocala	51,383	69,030	122,488	194,707	279.3
Leon	Tallahassee	74,225	103,047	148,655	192,493	159.5
Alachua	Gainesville	74,070	104,764	151,348	181,596	145.3
Lake	Leesburg	57,383	69,305	104,870	152,104	163.3
Collier	Naples	15,753	38,040	85,791	152,099	863.4
St. Lucie	Ft. Pierce	39,294	50,836	87,182	150,171	284.3
Okaloosa	Ft. Walton Beach	61,175	109,920	143,377	126,994	134.8
Bay	Panama City	67,131	75,243	97,740	126,994	89.1
Charlotte	Port Charlotte	12,594	27,559	59,115	110,975	781.0

Table 3. Population Growth in Florida's 25 Largest Cities, 1960–1990[20] (Ranked by 1990 Population).

City	County	1960	1970	1980	1990	% Change 1960–1990
Jacksonville	Duval	201,030	503,024[21]	540,898	635,230	217.1
Miami	Dade	291,688	334,859	346,931	358,548	22.9
Tampa	Hillsborough	274,970	277,753	271,523	280,015	1.8
St. Petersburg	Pinellas	181,298	216,159	236,893	238,629	31.7
Hialeah	Dade	66,972	102,452	145,204	188,044	180.7
Orlando	Orange	88,135	99,006	128,394	164,693	87.0
Ft. Lauderdale	Broward	83,648	139,590	153,256	149,377	76.1
Tallahassee	Leon	48,174	72,624	81,548	124,773	159.6
Hollywood	Broward	34,135	106,873	117,188	121,697	157.5
Clearwater	Pinellas	34,392	52,074	85,450	98,784	189.4
Miami Beach	Dade	63,145	87,072	96,298	92,639	46.8
Kendall (CDP)	Dade	—	—	41,100	87,271	—
Gainesville	Alachua	29,701	64,510	81,371	84,770	185.5
Coral Springs	Broward	—	1,489	37,359	79,443	—
Cape Coral	Lee	—	11,470	32,103	74,991	—
Pompano Beach	Broward	19,999	38,587	52,618	72,411	—
Lakeland	Polk	41,350	42,803	47,406	70,576	71.1
W. Palm Beach	Palm Beach	56,208	57,575	62,643	67,643	20.4
Plantation	Broward	4,772	23,523	48,501	66,692	1,297.6
Largo	Pinellas	5,302	22,031	63,800	65,674	1,139.4
Pembroke Pines	Broward	1,429	15,496	35,776	65,452	—
Sunrise	Broward	—	7,403	39,681	64,407	—
Palm Bay	Brevard	2,808	7,176	18,560	62,632	—
Daytona Beach	Volusia	37,392	45,327	54,176	61,921	65.2
Boca Raton	Palm Beach	6,921	29,538	49,505	61,492	—

9. Florida Administrative Code R9J-5.006.
10. Edward Montanaro, "Funding Growth Management: The Shape of Things to Come," in *Florida Environmental and Urban Issues,* Vol XIV, No. 1, October 1986, pp. 20–23.
11. John M. DeGrove and Nancy E. Stroud, "New Development and Future Trends in Local Government Comprehensive Planning," in *Stetson Law Review,* Vol. XVII, No. 3, p. 587.
12. Ibid., p. 577.
13. Westi Jo Dehaven-Smith and Robert Patterson, "The 1986 Glitch Bill — Missing Links in Growth Management," in *Florida Environmental and Urban Issues,* Vol. XIV, No. 1, October 1986, pp. 4–9.
14. This is the approach used by John M. DeGrove in his excellent analysis of growth management in Florida, California, Hawaii, Oregon, Colorado, and North Carolina. See DeGrove, *Land Growth and Politics,* op. cit. Note 1.
15. Nationally, of 53 graduate programs in urban and regional planning recognized by the Planning Accreditation Board in 1985, 22 had been initiated after 1970. Source: *Guide to Graduate Education in Urban and Regional Planning,* Fifth Edition, 1986, Association of Collegiate Schools of Planning.
16. The Section 702 program of the U.S. Housing Act of 1954 as amended provided funding for local governments to prepare comprehensive plans and related studies. Appropriations reached a high of $100 million in 1974 and 1975, but this program was zero funded and abolished by the Reagan administration in 1981. The Community Development Block Grant Program provided monies for local governments to conduct planning activities and between 1987 and 1988; 10–15% of annual allocations went for that purpose. However, this program, which was funded in the amount of $4.6 billion in 1980, was cut to only $2.9 billion in FY 1989. See Carl Feiss, "The Foundations of Federal Planning Assistance: A Personal Account of the 701 Program," in *Journal of the American Planning Association,* Spring 1985, pp. 175–184; Mary Nenno "H/CD After Reagan: A New Cycle of Policies and Partners," in *Journal of Housing,* March/April 1989, pp. 75–82.
17. Those planners who don't go along with elected officials do in fact run the risk of being terminated. See Nancy Wilstach, Kent Faulk, and Lou Ann Ray, "The Trials of Connie Cooper," in *Planning,* November 1990, pp. 12–15.
18. In April 1993 the Florida Legislature passed the Environmental Land Management Study Committee III bill (HB2315), significantly modifying the 1985 legislation. The Florida Chapter of the American Planning Association (APA) went on record expressing concerns about some aspects of this legislation, especially the proposed revamping of the state planning process through establishment of a strategic growth and development plan

and a phaseout of the Development of Regional Impact (DRI) process. The Florida APA felt that these actions weakened growth management (*Florida Planning*, Vol. V, No. 4, p. 1). This revised legislation could have the effect of weakening public planners' resolve to prepare the best possible plans and instead cater even more to the perceived developer-influenced agenda of elected and appointed officials.

19. Source: U.S. Census.
20. Includes incorporated municipalities and Census Designated Places (CDP).
21. Jacksonville and Duval County consolidated in 1967. Four small municipalities with a total of 20,000 people were left out of the consolidation.

2 The Evolution of Land Development Policy in Florida, 1850–1970

Florida was first described as an awesome subtropical paradise as early as the eighteenth century. The Philadelphia naturalist William Bartram wrote of his travels in Florida: "enchanting forests of live oaks, wild orange trees, towering royal palms and magnolias, water turning with trout of as much as 15, 20, or even 30 pounds, enormous alligators of up to 20 feet long and a profusion of other animals and birds including deer, bear, wolves, panthers, sandhill cranes, and wild turkeys."[1] In the early 1840s, Stephen R. Mallory of Key West, in describing the pineland ridge in south Florida between the Atlantic Ocean and the Everglades, stated "This is a fine country for a man who wishes to be independent. The woods and streams abound with game and fish, frost is rarely seen.... The most indolent man I ever knew prospered here."[2]

Despite its beauty and relative ease in liveability, Florida was actually the last state in our nation to become fully settled. In the peninsula, because of the flat terrain, torrential summer and fall rains left large areas of water that would stand for weeks before draining away or evaporating. The vast area of marshes and interior swamps were generally unfit for human habitation and were much too dense to facilitate transportation and communication. The salt marshes could be breeding grounds for swarms of mosquitoes, which could generate malaria and yellow fever. When Florida was admitted to the Union in 1845, it contained only 60,000 residents, virtually all of whom were confined to the area north of Gainesville, including Northeast Florida and the Panhandle region. Even by 1860, when Florida's population reached 140,000, settlement had progressed no further south than Ocala.

Development in Florida has proceeded from the beginning of statehood under an umbrella of political culture that can best be described as a mix of southern traditionalism and frontier individualism, as alluded to earlier by Stephen Mallory. When one looks at the state's history, we find that development up until 1970 was left almost exclusively to the private sector, without regulation or even guidance from government. In fact, the famous scholar V.O. Key notes in his classic *"Southern Politics in State and Nation"* that up to the end of World War II, government and politics in Florida could best be described as "Every Man For Himself." Stated Professor Key:

> Florida ranks high in political atomization. In it's [*sic*] politics, it is almost literally every candidate for himself. Ordinarily each candidate for county office runs without collaboration with other local candidates. He hesitates to become publicly committed in contests for state office lest he fall heir to all the local enemies of the statewide candidate. Each candidate for the half dozen or so minor elective state offices tends to his own knitting and recruits his own following. Senators and Representatives hoe their own row and each of the numerous candidates for governor does likewise.... Few politicians exert real influence outside their own county and those who can deliver their home county are few. Florida is not only unbossed, it is also unled.[3]

When we examine the history of Florida's development, not only does one find that government failed to provide leadership and direction for growth, but it actually assisted the developers in many different ways to maximize profits at the expense of taxpayers. Understanding this history is important in any evaluation of the success and/or failures of the 1975 Local Government Comprehensive Planning Act and the 1985 Growth Management Act. Despite the best intentions of the framers of these legislative packages, one cannot wipe out over 120 years of precedent with the stroke of a pen.

This chapter looks at state development policy evolution, including the emergence of major issues in distinct periods — 1850 to 1900, 1901 to 1945, the end of World War II, 1946 to 1970, and 1971 to 1975 — during which time Florida government moved 180 degrees, from a stance of laissez-faire to mandated planning for all units of local government.

1850–1900: Statehood, Civil War, Reconstruction, and the Railroads

In 1821, Florida was acquired by the United States from Spain. The territory was largely unsettled and was mostly owned by the Seminole

Indians. The Seminoles — about 5,000 strong — were dispensed by brute force in a series of wars that spanned over two decades, with the decisive battle taking place in 1857. Statehood did not come until 1845, after all other territories east of the Mississippi river with the exception of Wisconsin had become states. The 1830 U.S. Census found only 35,000 permanent residents. This population increased to 55,000 in 1840 and 87,500 in 1850 (see Table 1). According to the U.S. Census Bureau, up until 1850 Florida's population was exclusively rural. The growth between 1830 and 1850 was due mainly to the influx of two distinct groups. The first consisted of tens of thousands of poor white subsistence farmers and their families from rural Georgia, Alabama, and South Carolina who migrated south to the greener pastures of Florida's Panhandle, North Central, and Northeast Florida areas. Known as "Crackers" because of the long whips used to drive cattle, these farmers were, for the most part, rugged individualists, distrustful of government at any level, and people who fiercely valued their independence. The second group was made up of hundreds of more affluent whites, some from Georgia, Alabama, and the Carolinas, others directly from the British Isles or France. These whites opened plantations in Northeast and North Central Florida worked by tens of thousands of slaves, so that by 1850 Florida's population was almost 50% black. So in the years immediately following the onset of statehood, Florida's population of 80,000–90,000 was spread thinly between Pensacola and Jacksonville, in counties adjacent to the Georgia and Alabama borders south to Gainesville and St. Augustine. Governance was loose and widely decentralized, as it took days to travel between Jacksonville, Tallahassee, and Pensacola, which were only accessible by navigable waterway and rough overland trails.[4]

Florida as a frontier state was shaped by men who were preoccupied with growth and development. Because Florida's natural environment was complex and relatively little understood, it was impossible for early developers to fully contemplate the impact of their actions on the state's fragile wetlands, coastal mangroves, and the Everglades. Because the state's northern tier was mostly in hardwoods and elevated flatlands which could accommodate development in the form of cotton and tobacco plantations or subsistence farms, early settlers felt that the portion south of Ocala, St. Augustine, and Cedar Key could be made ready for development simply by clearing and drainage. The failure to appreciate and better understand

Florida's natural environment in areas that would eventually be settled after the Civil War was partly a reflection of the esthetics and ethical sensibilities of the times, which was best put by the Ian McHarg saying *"multiply and subdue the earth."*[5] Dredging and filling in the coastal estuaries and drainage of the Everglades would in time bring major difficulties to Florida, but in 1850 and up to the early 1900s few Florida officials and citizens had a premonition of problems to come.[6]

As land was taken from the Seminoles by U.S. troops and state militia, it reverted to federal ownership, which of course was eventually transferred to the state, or to private citizens either by grants in exchange for needed public improvements, such as railroads, or by purchase for plantations or homesteads. Two major factors initially shaped this transfer of land, the first being basic errors by surveyors, and the second being the Swamp Acts of 1850.

Surveyors' Errors

At the coming of statehood, almost four-fifths of Florida was still in public land owned by the federal government. By the end of the 19th century, several million acres of the federal domain had been disposed of by sales, grants to homesteaders, and set-asides for national forests and wildlife refuges. But about 24 million acres, or 65% of Florida, had been given to the state.[7]

The federal land grants to the state included considerable areas of submerged or "sovereignty" lands. Such lands are held in trust for the people by the sovereign or government with title to them automatically vested in the state once statehood was attained. Surveying these submerged lands was impossible given 19th century conditions. These lands included all bottoms of navigable lakes and all tidal bottoms, and about 9,000 miles of detailed shoreline, which is longer by far than that of any other state except Alaska. While a state may sell or permit the alteration of some limited part of its sovereignty lands, these parcels must in general be kept open to the public for fishing, navigation, and other purposes. Federal surveyors were assigned to prepare maps and charts on which the coastal shorelines and the shores of navigable rivers and lakes would be shown. All tidal waters were considered as either potentially or in fact navigable, along with any lake of 25 acres or more. Without modern surveying aids such as aircraft, motorboats, helicopters, and computer-aided compilation and analysis techniques, the 19th century surveyors missed hundreds of lakes and millions of acres of tidal bottoms, and these errors meant that lands were lost forever to public ownership. When boundary lines were mistakenly drawn below mean high water, sometimes well offshore, the boundaries described in deeds covering adjacent uplands actually took in sovereignty lands, including beaches.[8]

Authors describing growth management in states such as California and Oregon have taken pains to inform the reader that the ocean beaches in those states are almost always considered public lands. In addition, *access* to these lands is considered a major public issue.[9] It goes without saying that in the 19th century, especially in the period of 1830–1890, it was easier to survey the waters just off the California and Oregon coasts, with their relatively cool climate, than Florida, with its heat, humidity, mosquitoes, snakes, and alligators. Given Florida's individualistic population and relatively loose system of governance, coupled with the loss of ownership of beaches and navigable bodies of water, it is understandable why public access to the beach and other shoreline lands has never developed as an issue in Florida to the same extent as in states such as California and Oregon.

The Swamp Acts of 1850

While there was abundant, naturally well-drained land to support Florida's 87,500 residents in 1850, state officials were at that time contemplating draining the Everglades for agricultural purposes. In 1850, Congress passed the Swamp Lands Act, which allowed the state of Florida to claim "swamp and overflowed lands" for the purpose of making them productive for agriculture by drainage and the construction of levees. Florida claimed just over 24 million acres, with most of these lands located south of Ocala and within 50 miles of the Atlantic Ocean, Gulf of Mexico, or the western edge of Lake Okeechobee. Most of this acreage was made up of inland marsh and swamps, although it included some sovereignty lands missed by the surveyors and some parcels that were actually high and dry but included due to fraud.[10] The lands transferred to the state of Florida from the federal domain by the Swamp Lands Act and related statutes exceeded by several million acres that which was received by the largest western states, such as California and Montana.

The Swamp Lands Act was eagerly pushed by Florida's congressional delegation, as these legislators assumed it would be easy and profitable to reclaim these lands. Florida Senator J.P. Westcott, Jr., arranged for the federal government to commission a feasibility study of Everglades drainage. This study, made by Buckingham Smith, an attorney out of St. Augustine, reported in 1848 that the waters of the Everglades basin were in excess of 12 feet above sea level, and that these lands could be drained by making canals of streams such as the Hillsborough and Miami rivers and by giving Lake Okeechobee several outlets to the Gulf of Mexico and the Atlantic Ocean.[11] Smith, although conceding that the Everglades was a beautifully awesome and romantic sight, stated, "But if the visitor is a man of practical utilitarian bent of thought, the first and abiding impression is the utter worthlessness (of the region) to civilized man."[12] Smith felt that drainage of the Everglades could make possible the cultivation of tropical fruits such as guava and mangoes — products that could be grown nowhere else in the United States — and the production of basic staples such as sugar. General William S. Harney, who was in the Everglades during the Seminole wars in the 1840s, noted that "millions of acres of highly valuable land could be reclaimed from the region which otherwise could remain a wasteland fit only for the resort of reptiles."[13] Harney also felt that success in tropical agriculture once the Everglades was drained would make the United States independent of the West Indies for these products, protect the southernmost coast of the United States, and lead to the settlement of at least 100,000 people in the Everglades within five years."[14] Stated Harney, "Our coast in south Florida is now extremely exposed in time of war. This population would protect it ... and tend to the security of the entire southern portion of the Union in an eminent degree."[15]

Before actual drainage could occur, the Civil War, and then Reconstruction, took place. It was not until the 1880s that drainage would be initiated. However, what is important to remember is that almost at the same time as the onset of statehood, Florida's public and private leadership had committed itself to the mind-set that development at any price was a positive step forward, while ignoring any possible damage to the environment. This was to be the essence of public policy regarding land development in Florida until the 1960s, as well as elsewhere in the United States.[16]

The Coming of the Railroads

In 1851 the Florida legislature established the Internal Improvement Board. The purpose of this body was to manage all of the state's acreage, especially the huge amount of swamp and overflow lands. The legislature envisioned this board as a body that would encourage development of the state's lands primarily for agriculture and mineral extraction, which in turn would be facilitated by the construction of railroads and canals. The Internal Improvement Board was made up of the governor and four of his appointees, including two railroad presidents, the state comptroller, the treasurer, the secretary of agriculture, and the registrar of state lands.

The board's mandate included the building of railroads through land grants and the floating of bonds to cover construction costs, including site preparation, rail, bridges, trestles, and rolling stock. In land grants, not only did the railroad companies receive a 200-foot right of way, but for a distance of six miles out from either side of the track they received alternate sections of land (640 acres each) in a checkerboard pattern on either side of the rights of way. The lands were to be paid off from receipts accruing to the railroads, but if these receipts were not forthcoming, bondholders would receive state lands that were pledged against these obligations. Therefore, these bonds were a combination of revenue and general obligation issues that became a "win-win" scenario for the bondholders.[17]

Under this arrangement, several railroad companies built and/or improved lines in north Florida prior to 1861. At the onset of the Civil War, railroad lines in Florida extended from Fernandina west to Tallahassee, and southwest to Cedar Key by way of Gainesville. However, the Civil War ruined these companies, and their property reverted to the state. By 1870, the Internal Improvement Board was short of almost one million dollars in cash, could not meet its rail bond obligations, and was forced to fold all operations.[18]

Reconstruction ended in 1877, and by 1880 the State of Florida was facing a public land crisis because of its million dollar debt. But Governor William D. Bloxham, a Democrat elected shortly after Reconstruction, found a way to solvency. He convinced a wealthy young Philadelphia saw manufacturer named Hamilton Disston to buy four million acres of swamp lands owned by the state for $1 million, or 25 cents per acre. Disston, who had visited Florida on several occasions on fishing and hunting trips, realized the potential of these lands, and his cash allowed the Internal Improvement Board to pay off its rail bond obligations just before one creditor took steps to obtain a judgment of 14 million acres in exchange for his coupons. Disston's holdings were mostly in 10,000-acre tracts stretching from Duval County (Jacksonville) to Lake Okeechobee along the Atlantic Ocean. Just as the state had selected tracts that due to survey errors were neither swampy nor overflowed, Disston's own selection included several upland tracts. In addition to his initial four million acres, Disston received a bonus — title to one-half of all of the swamp land he could reclaim from the upper Kissimmee basin down through the Everglades. Altogether almost nine million acres were to be reclaimed. This "deal" made Disston the largest landowner in Florida, and possibly the United States as well.[19]

Luther Carter details Disston's dredging operation as follows:

> A Disston dredge worked its way from Fort Myers up the Caloosahatchee River, a stream then rising in the marshes and shallow lakes east of Lake Okeechobee. Finally, cutting a canal from the upper searches of the Caloosahatchee to Okeechobee, the dredges also made the meandering Kissimmee River more navigable, but without greatly altering it and they cut canals between all major lakes of the Kissimmee basin, thus opening up the basin to steamboat traffic as well as draining some of it's [*sic*] marshes. An attempt was made to dig a canal from Lake Okeechobee south across the Everglades to the Shark River but after cutting less than 10 miles through the deep muck, the dredge encountered hard rock and could go no further.[20]

In 1885, the old Internal Improvement Board was replaced with a new body — the Board of Trustees of the Internal Improvement Fund. Under the new state constitution, the board consisted of the governor and the Florida Cabinet, along with several independently elected officials. With the new board's return to solvency after the Disston sale, the state was now able to resume its pre–Civil War practice of land grants to railroad companies. The period from the end of reconstruction to the turn of the century is generally regarded as Florida's "Bourbon" era. The Bourbons were Democratic leaders who restored the political supremacy of well-to-do propertied white Floridians, including the remnants and descendants of the pre–Civil War ruling class. By the use of literacy tests, the poll tax, outright violence, and economic intimidation, Florida's 200,000 black citizens — most of whom were freed slaves and their children — were forced out of minor elected offices, barred from the polling place, and reduced to a position of peonage during the period of 1880–1900. The poll tax and literacy tests also served to exclude poor whites, many of whom served in the Confederate Army, from participating in the electoral process. By 1885, the Bourbons had an ironclad grip on Florida's statewide and local politics and were free to make deals as they saw fit without the benefit (or inconvenience) of public scrutiny.

The restoration of the Bourbons was accomplished largely through an alliance with newcomers — mainly northerners — who had money to invest, such as Disston and the major railroad builders. What is noteworthy, as the case studies in this book well show, *this old alliance continues to the present time,* albeit in a more complicated format. Charlton W. Tebeau in his *A History of Florida* comments on this phenomenon:

> The Bourbon blueprint for state development was fairly simple. Government activity and costs had to be kept at a minimum and taxes correspondingly low with no incubus of state debt. Government encouragement but never restraint was offered to those who would invest their money and look for their futures in the state. Bourbons maintained an image of economy and honesty in contrast to the previous [Reconstruction] regime [but] could be accused of extravagance...in the wanton fashion in which they offered the state's natural resources for development and exploitation.[21]

By the end of the century, the legislature and the reconstituted Board of Trustees of the Internal Improvement Fund had given the railroad companies about nine million acres — almost one-fourth of all land in Florida. Because of inaccurate surveys and the Board's failure to hold sovereignty lands for the public, in many cases tidal lands were deeded to railroads as swamp and overflow lands. For example, a deed issued by the Board on February 8, 1898 conveyed 14,000 acres of swamp and overflow land in the Florida Keys to the Jacksonville, Tampa, and Key West Railway as part of about 550,000 acres due this company under a grant by the legislature, which was conditional upon laying 55 miles of track in north Florida. But the land granted was, by the terms of the act, supposed to be lands "lying nearest the line of said road." The acreages specified in the deed were not swamp and overflowed lands by any reasonable definition, but instead tidal or sovereignty lands in the Florida Keys, hundreds of miles away from the track to be laid.[22]

Florida's lavish land grants provided incentive for holders of investment capital to put their money into the construction and expansion of the state's railroad system. Two such capitalists were Henry B. Plant and Henry M. Flagler. Plant, who was principal owner of the Southern Express Company, constructed a railroad west from Daytona Beach past Orlando to Tampa. This railroad was completed in 1884. At the terminus of his railroad, Plant built a luxury hotel, which is now the principal building of the University of Tampa just outside this city's downtown. Plant's railroad created an economic boom for Tampa, whose population rose from only 720 in 1880 to 5,532 in 1890 and 15,839 in 1900. Flagler, a former associate of billionaire John D. Rockefeller in the Standard Oil Company, constructed his Florida East Coast Railway south from St. Augustine beginning in the early 1890s. By 1898, this railway reached Miami. It cost $30 million to construct, but once completed, opened not only Miami and South Florida for transportation of tropical and citrus agricultural products to northern markets (the refrigerated boxcar was perfected by 1890, thus making it possible to ship fresh fruit over a two-day journey to marketplaces), but tourism, retirement, and leisure as well. Miami's population, only 116 in 1880, rose to 1,681 by 1900.

By 1900, Florida's population, only 269,000 in 1880, had almost doubled to 529,000. Over 20% of the 1900 population was considered as "urban" by the U.S. Census Bureau. Due to the extraordinarily generous land grants by the state, Florida's basic railroad network of today was in place. The population previously confined to the northern part of the state was now free to expand to almost all areas. Miami and Tampa now had rail connections to the rest of the United States, which meant that now citrus, winter vegetables — and in Tampa, phosphate — could be shipped to national and international markets. But growth was just beginning to come. During the period of 1900–1945, growth would intensify. Much of the Everglades would be drained for agriculture. Tourists would now find their way to Miami Beach, Daytona Beach, and other resort areas. And although there were some significant attempts to curb land giveaways by the state to private interests, there was no corresponding attempt to manage growth. The swamps, with their estuaries, cypress heads, and water recharge areas, were seen as obstacles to progress. Growth was considered a good to be achieved at any cost. The words "preservation" and "ecology" were as foreign as a language from Mars or elsewhere in outer space.

1900–1945: The Progressive Era, Drainage, Urban Land Boom, the Depression, and World War II

Impact of the Progressive Era

By the turn of the century, Florida's population exceeded one-half million permanent residents, 20% of which were classified by the U.S. Census Bureau as "urban."[23] At that time, the Progressive Era was upon America, and its impact was felt in Florida. The old Bourbons were gradually replaced during the 1890s by new business interests who were a mix of the children of landholding Bourbons and an emerging management class representing the railroads, timber and phosphate companies, agricultural processors, and wholesale traders. This group's position was essentially that in order for Florida to prosper, new capital must be attracted, taxes must remain low, and a minimum of governmental regulation should be maintained. On the other hand, a new group of "urban" interests, including retail merchants, small businessmen, and a growing professional class of physicians, attorneys, engineers, and accountants, attempted to push another agenda. This alternative group argued that state lands should be made available to actual settlers rather than large investors, that more state services should be provided in the areas of health and education, and that opportunities for individual entrepreneurship should be pushed. However, these two camps coexisted within the Democratic Party, and their agendas were similar in that both rejected participation in the state's political process and economic opportunity the working class, poor, and of course, Negroes.[24]

Supported by both elements of this new coalition of interests, William Sherman Jennings was elected governor in 1900. Scholars agree that while Jennings was competent in the area of public finance, his major contribution was the administration of state lands.[25] Because, since 1881, the state had granted more land to railroads than it held or was likely to receive from the federal government, these railroads protested sales of any major state lands until their claims were resolved. Jennings, after careful study by independent consultants, decided that the state should drain and manage state lands in the interests of settlers, and moved the trustees of the Internal Improvement Fund to sell, in 1902, 100,000 acres of swamp and overflow lands in north Florida for $225,000 to satisfy state financial obligations and drain more land. The railroads sued, but in 1907 the U.S. District Court ruled that the State of Florida had the right to dispose of swamp and overflow lands for the purpose of draining as a priority over encouraging the construction of railroads and canals.[26] When the state received 2.8 million acres of swamp and overflow land from the federal government in 1903, Jennings refused to surrender any of it to railroad interests and instead provided for its eventual drainage and sale to developers and settlers. The railroads again sued, but the U.S. District Court's decision of 1907 invalidated their claims.[27]

Another instance of the Progressives' impact on Florida politics was that the law that provided for the direct primary to nominate candidates for office was strengthened. Starting in 1902, all U.S. congressmen and state officials were nominated in a statewide primary election, with run-offs if no candidate received a majority of the vote. This primary replaced the old party conventions and provided for greater accountability, as candidates now had to appeal to a broader spectrum of the population than beforehand. By the end of Governor Jennings' term, state policy regarding land disposition had shifted from railroad giveaways to sales to individuals for drainage and reclamation, either for their own use or subsequent sale to farmers and ranchers.[28] Thus Florida's public policy regarding land development advanced one step further.

Drainage

In 1904, Napoleon Bonaparte Broward was elected governor of Florida at the relatively young age of 47. A Progressive Era Democrat like his predecessor Jennings, Broward promised to carry forth the former governor's policies regarding land development: drainage of state-owned swamps and overflow lands, then sale for farming, ranching, and timber cultivation.[29] Broward presented himself as the "people's candidate." He promoted himself as a candidate not tied to the railroads and their land agents, and promised that if elected he would, as chair of the Trustee Board of the Internal Improvement Fund, deny their claims. He sought support from small businessmen, farmers, and ranchers. His most effective campaign slogan was "Water will run downhill," a clear but simple endorsement for drainage.[30] Campaigning on this and related themes, Broward won the Democratic primary, then the general election.

Actually, former Governor Jennings had started the drainage ball rolling in 1903 when he hired a U.S. Department of Agriculture engineer to survey lands in the Miami area. The surveyor, Charles G. Elliott, found that dredging to a grade of 0.3 to 0.4 feet per mile to the center of the area would facilitate drainage. In 1905, the legislature passed a comprehensive drainage law that provided for a Board of Drainage Commissioners empowered to construct a canal system that would drain and reclaim swamp and overflowed lands in the state. Broward appointed his predecessor, Jennings, as general counsel to this new board. After the federal courts upheld Broward's contention that the state's right to drain and dispose of lands by sale was superior to the railroads' claims, drainage was ready to proceed. Surveys were completed in 1908. Even prior to that, the Board of Drainage Commissioners had created an Everglades Drainage District of 4.3 million acres located east and south of Lake Okeechobee. More work was to begin. In order to build the canals necessary for drainage, the state had to resort to land sales. Sales were no problem, since with drainage now a real possibility — as with Disston's efforts in the 1880s — new technology facilitated a land boom. For example, in June 1908, a Colorado based firm purchased 27,500 acres at $2 per acre. In October of that same year Walter R. Comfort of New York purchased 6,400 acres; in November, J.H. Tatum of Miami bought 12,000 acres and in December, the Davie Realty Company purchased 80,000 acres and later part of that land became the town of Davie in Broward County.[31]

The most significant purchase was that of Richard J. Bolles. Bolles, born in New York, is described by Luther Carter as a self-styled hustler — "one of the earliest and most innovative salesmen of Florida swamp land."[32] Bolles' purchase in December 1908 was 500,000 acres for $1 million, or $2 per acre. $500,000 of Bolles' purchase price was to be used exclusively for drainage, and payments were to extend through 1916. Bolles subdivided his holdings largely into 10- and 20-acre farms and sold each one for $20 per acre.[33] By 1911, Bolles' Everglades Land Sales Company was sending out literature claiming that 224,000 acres had already been sold and only a limited acreage was left at $50 per acre. The sales were made by smooth-tongued slicksters who toured the Midwest giving slide shows and providing refreshments. Alfred Jackson Hanna and Kathryn Abbey Hanna describe Bolles' operations as follows:

> The Florida Fruit Lands Company, first of the Bolles companies to be organized, introduced the contract method of land sales. Its holdings of 180,000 acres in Dade and Palm Beach counties were divided into 12,000 farms...8,000 farm tracts were 10 acres in extent.... Within the holdings, a town site called Progresso was designed by the company; it was to consist of 12,000 lots in addition to reservations for streets, factories, churches, schools, parks, public squares, and public buildings. Twelve thousand contracts were put on sale at $240 each for which payment could be made at the rate of $10 a month. Each contract holder owned a farm the size and location of which was unknown (at the time of sale) and a town lot.... Glib salesmen toured the middle west aglow with enthusiasm over the Everglades "Tropical Paradise." They called it the "Promised Land," the "Poor Man's Paradise," the "Land of Destiny," and the "Magnet Whose Climate and Agriculture Would Bring the Human Flood." There life was full and riches were to be had on every hand, or at least a generous competence for old age. They conclusively proved that a 10 acre farm stocked with cows, hogs, and chickens could maintain a family and provide extras as well.[34]

Much of this land sales activity was speculative in nature. Tebeau notes that at least 60% of the purchasers bought only to hold for a higher price later. Many purchasers defaulted on their land contracts, especially during the recession of 1910–1911. Other purchasers found that when they travelled to their "farms" they found them to be under water.[35] By 1911, opposition to Bolles' land sales began to mount, with Progressives demanding a congressional investigation. The investigation took place in February 1912, and although the results were inconclusive, the publicity surrounding it caused a tremendous drop in land sales. Some promoters were tried for fraud, and in November 1914 a Kansas City promoter who had sold 20,000 people 50,000 acres of Everglades land in the form of 10- and 20-acre farms was convicted. Bolles himself was indicted on fraud charges in 1913, but these charges were dismissed and he died before new charges could be brought.[36] During the trial of the Kansas City promoter, the governor of Florida, Park Trammell, appearing as a defense witness, was forced to admit that Florida's only means of raising funds for drainage was the sale of state lands and that the purchasers of these lands like Bolles had brought on federal investigations and bad publicity, which in turn stopped further sales, "very seriously crippling the drainage fund." The great majority of people who by 1912 had purchased land eventually forfeited it either by defaulting on their contracts or failing to pay taxes.[37] This phenomenon was to become a familiar pattern in Florida land sales that has lasted to the present time.[38]

By the early 1920s, the drainage program was proceeding once more, except by now large tracts of land were being sold to farming corporations. By 1921 there were about 2,000 inhabitants on or near Lake Okeechobee. The Florida East Coast Line extended its tracks to Okeechobee in 1915 and Belle Glade in 1926. The Atlantic Coast Line reached Moore Haven in 1918 and Clewiston in 1922. Florida's highway network, begun in 1911, reached the Glades area by 1920. By 1925, over $10,500,000 in lands for drainage had been sold. Canals constructed by that time included the Caloosahatchee Waterway between Lake Okeechobee and the Gulf; the St. Lucie Canal between the lake and the Atlantic; and the Miami, North New River, Hillsboro, and West Palm Beach canals, all between the South Shore of Lake Okeechobee and the Atlantic in a diagonal south-southeast pattern. By 1929, despite a devastating hurricane the year before that resulted in the loss of 1,800 lives, $18 million had been spent for drainage. Population in the drained land — by that time known as "the Glades" — had reached 50,000, and over 200,000 acres had been drained and 100,000 placed in cultivation. Some 586 miles of paved roads and 210 miles of railroads had been constructed. About 440 miles of canals were constructed, and land, then worth $300 million, had increased 60 times its value since 1905 and was being used to raise sugar cane, citrus, winter vegetables, and cattle.[39] By 1930, Florida was the nation's sixth largest agricultural producer in terms of dollar volume of crops, and the largest single area devoted to agriculture was in the Glades. The drainage program was a success by any measure, bringing wealth to the growers and good times for merchants — but field workers, especially poor blacks imported from North Florida, Georgia, Alabama, and the Carolinas to man the farms, did not share in this new prosperity.

The Urban Land Boom of 1915–1925

The U.S. Census of 1910 showed Florida's population at 752,619, a 42% increase from 1900. By 1920, the state's population had risen to

968,470, and by 1930 there were 1,468,211 residents in Florida. Not only did the state's population double in this 20-year period, as at least 700,000 new residents came to the state, but the percentage of the population identified as "urban" by the U.S. Census Bureau rose from 20% in 1910 to 52% in 1930. On the eve of the Great Depression, Florida was for the first time an urban state — one where the percentage of residents residing in urban areas outnumbered those living in rural districts.

There were several reasons for the rapid growth of residents living in urban areas during the period 1910 to 1930. First of all, the state's railroad network was virtually complete by the mid-1920s, which facilitated not only the movement of passengers in the form of new residents and tourists, but more so, freight operations, especially those that shipped agricultural products to processing centers (most likely to be located in an urban setting of 2,500 or more people), then to the northeastern and midwestern U.S. wholesale distribution points. Second, the boom economy of the 1920s encouraged tourism and the phenomenon of the "snowbird," or winter visitor. And most important, the first two factors gave rise to a third — the emergence of entrepreneurs who saw opportunities for urban development and made the most of them.

One of these enterprising entrepreneurs was Carl Fisher. A wealthy winter visitor to Miami beginning in 1910, Fisher had made his fortune selling bicycles and automobiles.[40] Fisher took advantage of the Riparian Act of 1855, by which the Florida legislature, in their mad rush to foster development, allowed owners of riparian property to fill the adjoining tidal land out to the channel for construction of warehouses, wharves, piers, and related facilities. A later Riparian Act in 1921 declared as a policy matter that the state benefits from waterfront development, and stated explicitly that warehouses, dwellings, and other buildings could be constructed on this filled land.[41]

Armed with the legal authority of the state and newly perfected technology, Fisher used dredging and filling procedures to first develop the Point View subdivision on Miami's bayfront in 1912. In 1913 he received 200 acres near the south end of what was to be Miami Beach, in partial payment of a $500,000 loan to John Collins, who needed the money to finish construction of a bridge across Biscayne Bay to the Atlantic Ocean beach. Over a 15-year period beginning in 1913, Fisher was able to fill in the one-mile-wide mangrove swamps, obtaining without difficulty permits from the Internal Improvement Board and the U.S. Army Corps of Engineers. Fisher was keenly aware that filling in the mangroves would triple the usable land area and eliminate breeding grounds for mosquitoes and sandflies, and that dredging would provide enough depth for yacht harbors. Polly Redford, in *Billion Dollar Sandbar,* provides a description of Fisher's dredge and fill operation as follows:

> In the summer of 1913 new lands began rising from the bay at the south end of the beach where Fisher...spent $600,000 replacing a thousand acres of mangroves with six million cubic yards of fill. The forest was cleared by gangs of Negro laborers.
>
> ...They worked hip deep in mud, a pall of smoke hanging over them because smudge pots and bonfires of palmetto fiber were the only defense against clouds of mosquitoes and sandflies that made life difficult for man and mules alike. When the many branched mangrove roots proved ruinously expensive to remove, they were cut off to cover later on. At the water's edge, steam shovels heaped the bay bottom into dikes while pile drivers sank rows of support for a bulkheaded shoreline of wooden timbers anchored to pilings with steel cables. The shoreline alone cost $10 a running foot. Behind it, long pipelines reached from the mangrove stumps to the bay where suction dredges burrowed into the bottom, turning the water muddy milk white. Fisher's largest dredge, the Norman H. Davis, could pump fill through a mile of pipe and in places where this was not far enough, another dredge was rigged in tandem to boost the pressure...
>
> So year by year, a uniform five-foot plateau spread northward along the bayfront. As it rose, the bay bottom fell, and what had been hundreds of acres of turtlegrass flats covered with a foot or two of clear water became a deep, turbid pool running parallel to a smoothly bulkheaded shore. In this manner, the original landscape was erased as if it had never been and a more saleable one built in its place.[42]

So, Miami Beach was built as a result of this operation. From nothing in 1910, Miami Beach's 1920 population of 644 residents skyrocketed to 9,211 in 1930. Fisher and his associates were making land sales in the range of $20–25 million every year between 1920 and 1925. The dredge and fill techniques improved over time. Soon the steam shovels would be replaced by mammoth diesel-powered machines, and the wooden timbers replaced by steel channels; but the basic principles remain the same today.

Dredge and fill procedures transformed the Florida shoreline into a concrete and asphalt jungle of high rise hotels, condominiums, office buildings, and streets without regard to ecological and environmental damage. Polly Redford stated, "From Florida, the notion spread northward and eventually proved so profitable that today no marsh or mudbank between Boston and San Francisco is safe."[43]

Luther Carter notes that Fisher was aided in his push to destroy the mangroves for pure profit by state government. In 1913, the legislature in fact authorized the Internal Improvement Board to sell sovereignty lands and Fisher was actually allowed to buy the bay bottom on which Star Island was built in Biscayne Bay. This new island, built by dredge and fill methodology, was half a mile long and a quarter of a mile wide. Fisher paid the state only $17,000 for the bay bottom, and after bulkheading and filling it sold the land for $200 per front foot, for total sales of $15 million and a profit of at least $10 million. As a bonus, the state gave Fisher, at no cost, all of the fill material pumped from the bay bottom.[44] Carter's somewhat bitter commentary on Florida's government aid to developers such as Fisher, with limited benefits to average citizens and destruction of the shoreline environment, reads as follows:

> Whatever the merits of Miami Beach and its satellite islands as tourist attractions or commercial assets, north Biscayne Bay's beauty and extraordinary good fishing had been impaired. Moreover, for the sacrifice of the sovereignty lands that had gone into the making of this artificially created environment, the state had received virtually no return. But the state as represented by the legislature and the I.I. (Internal Improvement Fund) Board was aware of no sacrifice, for during the Fisher era and long afterward, neither state nor federal law recognized any public value in turtlegrass flats or mangrove swamps. Another half century would pass before the value of mangroves would be fully understood, even by marine biologists.[45]

With dredge and fill operations in full force, growth in Miami took off during the 1920s: assessed value of real property increased from $63.8 million in 1921 to $421.1 million in 1926. In the same period, the mileage of paved streets went from 32 to 420. Building permits rose from a value of $4.5 million annually to $63.8 million annually.[46] But growth was not limited to Miami and southeast Florida.

There was a recession in Tampa immediately after World War I, as two wartime shipyards closed, leaving 5,000 men unemployed. The cigar industry suffered a strike that idled 10,000 workers, and a severe hurricane in October 1921 added to the despair, with $150,000 in property losses, 15 dead, and over 200 injured. However, by 1922 the economy picked up as tourists — most of whom arrived by automobile — discovered Tampa and its warm winter weather and unhurried pace. Most filled their cars with bedding and canned goods for their journey. In the city's DeSoto Park these visitors organized themselves as the "Tin Can Tourists of the World."[47] Tampa's population increased from 15,839 in 1900 to 51,606 in 1920 and 108,391 in 1940 as a result of growth propelled first by railroads, which made manufacturing and shipping profitable, and then by tourism made possible by the development of Florida's highway system, which was linked nationally to cities in the Midwest and Northeast.

In 1926, the Gandy Bridge was completed, linking Tampa directly to St. Petersburg. Now there was a direct roadway connection across the Bay, and as a result, growth in St. Petersburg and other Pinellas County communities and the west side of Tampa Bay took off. Lots were platted and sold by developers even before the bridge was completed. As a result, St. Petersburg's population rose from 14,737 in 1920 to 60,812 in 1940, almost a fivefold increase.

The most spectacular development in Tampa during the 1920s was the creation of Davis Islands. D.P. "Doc" Davis had seen firsthand the development of low-lying islands and bay bottom land in Miami, where he had worked as a salesman for one of Carl Fisher's associates. He acquired Big and Little Grassy Keys, located in the mouth of the Hillsboro River, and organized a major development and sales campaign. Using dredge and fill methods perfected by Fisher in Miami, Davis filled in these islands, and the first block of 300 lots sold on October 4, 1924 — in three hours — with receipts totaling $1.7 million. The most astonishing thing was that on the day of this gigantic sale, almost all of the lots were under water. Before the end of 1925, the entire development had been sold out, for a total of $18.1 million.[48]

By 1926, the Florida urban land boom was coming to an end. Just like the Bolles swamp land sales in 1909–1912, there were numerous accounts of misrepresentation and outright fraud, and the resultant publicity caused potential buyers to back away. Many purchasers defaulted on their contracts and/or failed to pay taxes. The coup de grace was the hurricane that struck Miami on September 26, 1926. Total casualties were 372 dead, 6,281 injured, and 17,784 families left fully or partially homeless.[49] No sooner was the debris from this hurricane cleared away than the 1928 hurricane at Lake Okeechobee hit, with even greater loss of life and property damage. Then in 1929, Black Friday took place, wiping out the stock market and plunging the nation into the Great Depression.

Growth During the Depression and World War II

In 1930 Florida's population stood at 1,468,211 — a 52% increase from that in 1920 — and 51.7% was urban in character. By 1940, the population had risen to 1,897,414; almost 430,000 new residents had been added during the most economically disastrous period in our nation's history. Of that population, 55% lived in urban areas by 1940. Major cities reflected the statewide increase — Miami went from 110,000 in 1930 to 172,173 in 1940. Tampa grew from 92,500 to 108,391 and Jacksonville increased its population from 124,648 to 173,065.

State policy during the Depression was to keep taxes as low as possible. A constitutional prohibition on enactment of a state income tax in 1926 certainly did nothing to dissuade potential new residents from moving to the state. Taxes on horse and dog racing introduced during the depression helped to prop up state government in this period. Drainage operations continued in the Everglades, but at a much slower rate than that during the two previous decades. Tourism almost hit rock bottom, and many hotels in Miami, Miami Beach, Tampa, and St. Petersburg closed forever. However, newcomers with financial resources found a wide variety of affordable housing units available that were left fully or partially vacant as a result of the 1920s land boom.

The New Deal helped greatly to provide needed relief in Florida. Expenditures of federal funds in the state rose from $12,722,000 in 1930 to $62,718,000 in 1934, and averaged over $54,000,000 in the next three years. In 1938 the total dropped to $48,657,000 but rose to $67,218,000 and $64,920,000 over the next two years because of the "second depression" in 1938. By June 1938, the Public Works Administration had completed 137 projects, including 42 schools, 27 water supply facilities, and 6 sewer systems. By 1940, relief employment was being provided for almost

1,200 men, and the Civilian Conservation Corps enrolled 2,400 boys and employed 400 instructors and related personnel.[50]

A major New Deal project was the Cross Florida Barge Canal. As early as the founding of St. Augustine, explorers had dreamed of a waterway from the St. John's River to Tampa Bay, thus linking the Atlantic Ocean and the Gulf of Mexico. By the time Franklin D. Roosevelt had taken office as president, no less than 28 surveys of possible canal routes had been made. By 1934, the U.S. Army Corps of Engineers concluded that the best route was a stretch from the St. Johns and Oklawaha Rivers through a man-made channel passing through Ocala to the Withlacoochee River and into the Gulf at Port Inglis.[51] The project was approved by President Roosevelt in 1935 with an appropriation of $5 million to initiate work. The original appropriation was used up by the summer of 1936. Because of congressional opposition to the use of relief funds for this project, work stopped, and World War II began before it could be restarted.[52] The ultimate fate of this project will be discussed in Chapter 3.

One positive aspect of the Depression years was that finally, in 1939, the Florida legislature granted the right to local government to adopt zoning regulations. This took place 13 years after the Euclid v. Ambler Realty decision of 1926, which legitimized zoning. By 1940, Florida's population was 1.9 million — 55% of which was urban.

World War II found a huge expansion of military facilities in Florida for two reasons: (1) the belief that German submarines could attack the Atlantic Coast, and that the instability of nations in the West Indies — such as Cuba, Haiti, and the Dominican Republic — made the southern tip of Florida extremely vulnerable to attack; and (2) the warm and usually clear skies made for excellent flying and sailing conditions. During the war years over 1,600 miles of highway were either constructed or repaired,[53] over 400,000 troops were stationed in army, navy, and air corps facilities — many housed in hotels used for tourists in previous years. Like the rest of the country, there was full employment, with the state's unemployment insurance fund growing from $12 million in 1940 to $50 million in 1944. Florida had only six major airfields capable of handling all types of aircraft in 1940; by 1945 there were over 40 airfields of this type. After the war, most were disposed of for commercial airfields, such as what are now Tampa International Airport, Orlando International Airport, and facilities of similar magnitude and type in Jacksonville, Gainesville, West Palm Beach, and

Ft. Lauderdale. One former airfield in Boca Raton became the campus of Florida Atlantic University in 1964, and even today the old runways are used for parking lots.

The war had a greater impact on Florida than elsewhere in the nation. Florida's population increased from 1,897,414 in 1940 to 2,771,305 in 1950 — a 46% jump. In contrast, the entire Southeast grew only 14%, and the nation as a whole only 15%. In the peninsula portion of Florida, growth between 1940 and 1950 was a whopping 81.5%. But Florida's greatest period of growth was just ahead.

1945–1970: Continued Growth, but Problems Appear

Between 1945 and 1960, Florida's growth continued at a breakneck pace. By 1950 there were 2,771,305 residents, a 46.1% increase from that in 1940. Most of this growth was directed into the peninsula, especially around urban centers, such as Miami, Tampa, and St. Petersburg. In 1940, Miami boasted 172,172 residents, and that figure grew to 249,276 in 1950. Tampa's population increased during that same time period from 108,391 to 124,681, and St. Petersburg grew from 60,812 to 96,738.

Florida's population growth really took off from 1950 to 1960. By 1960, 4,951,560 people called Florida home — 78.7% more than in 1950. The growth of urban centers was remarkable during this period. While most central cities in the United States lost population during this decade, only one did in Florida — that being Jacksonville, which experienced a slight drop from 204,517 to 201,030. Miami grew from 249,276 to 291,688; Tampa from 124,681 to 274,970; St. Petersburg from 96,738 to 181,298; Orlando from 52,367 to 88,135; Ft. Lauderdale from 36,328 to 83,648; and West Palm Beach from 43,162 to 56,208.

This growth was spurred by several factors. First of all, Florida had been "discovered" by hundreds of thousands of veterans stationed there during World War II. When they arrived back home remembering the pleasant, sunny winters and the vibrant boomtown environment, thousands grabbed the opportunity to move there, taking new jobs made possible by the booming agribusiness and tourist industries, or at least took their vacations in Florida until jobs became available. The tourist industry picked up because not only were US 41 and US 1 four-laned for most of the journey from the Midwest and Northeast, respectively, by the early

1950s, thus facilitating automobile travel, but commercial airlines now outfitted with brand new jet aircraft were, by the early 1950s, able to offer discount airfares for a two- to three-hour trip to Florida cities. The Miami Airport had only 500,000 enplanements in 1940, but by 1960 that number reached 4.6 million — 4 million being domestic, while 600,000 were international. Florida boasted 5 million tourists in 1940, and that figure grew to 13 million in 1960, with $2 billion being spent by that group. While only 2% of all tourists arrived by commercial airlines in 1940, 23% did so in 1960.

Due to advances in medical science that prolonged lifespans and improvement in public and private sector pension plans, retirement helped to spur the Florida economy in the 1950s. Making this retirement boom work was the perfection of home air conditioning; first the "window units" in the 1950s, then beginning in the early 1960s, central air conditioning systems became available for even the most modest of homes. The phenomenon of air conditioning not only facilitated year-round living in this humid semitropical climate previously unbearable six months of the year, but it also created a summer tourist industry to complement the standard winter season. By the late 1960s, the interstate highway I-75 linked Florida with the Ontario Province in Canada and the midwestern states of Illinois, Indiana, Michigan, and Ohio. The northeastern megalopolises and Quebec were joined with Florida by I-95 except for a small stretch between Ft. Pierce and West Palm Beach. Tourism reached 18 million visitors annually by 1970, and by 1990, 25 million tourists were visiting Florida each year, spending a total of $10 billion annually.

The economic engines of agribusiness, tourism, and retirement/leisure were joined by manufacturing by the late 1950s. A decision was made in 1946 by the U.S. Air Force to locate a long-range missile flight testing center at what was to become Cape Canaveral. The site was first known as the Atlantic Missile Range, and the first space flight was made there in 1961 by Alan B. Shepard and, on February 20, 1962, Lt. Col. John Glenn became the first American to orbit the Earth, thus placing Cape Canaveral on the national and worldwide map. The development of the Atlantic Missile Range resulted in the incubation of a major aerospace industry located in a belt between Orlando and Melbourne, one that remains to this day. Another factor spurring industrial growth in Florida was the need of Floridians for consumer products unique to the state, such as aluminum screens for patios and pool enclosures, that could be manufactured and shipped to markets at a lower cost than similar products made elsewhere. By 1962, over 200,000 people were employed in Florida's factories, receiving wages totaling just under $1 billion, and manufacturing accounted for 13% of all Florida jobs.[54]

By the mid-1960s tourism had become Florida's number one source of income, followed by agriculture, retirement/leisure, and manufacturing. By 1970, Florida contained 6,791,418 residents, a 37.2% increase from the number in 1960. By 1970, 81.7% of Florida's population was classified by the U.S. Census Bureau as "urban," compared to only 56.5% in 1950. Urban growth was astonishing for both central cities and suburbs. Miami grew from 291,688 to 334,854 during this period. Tampa grew from 274,970 to only 277,753, which was still remarkable because virtually all U.S. cities of 250,000 or more in 1960 lost population by 1970. Other cities that posted strong growth rates during the 1960s were Ft. Lauderdale, 83,648 to 139,590; Hollywood, 34,135 to 106,873; and the Miami industrial suburb of Hialeah, 66,972 to 102,452. But, beginning in the 1960s, the real growth was in suburbs adjacent to major cities, as evidenced by total population growth in the counties. Dade grew from 935,047 to 1,267,792; Broward's population almost doubled from 333,964 to 620,100; Palm Beach grew from 228,106 to 348,993; and Pinellas jumped from 374,665 to 522,329. However, this growth — even as early as the late 1940s — was beginning to manifest some serious problems. These problems would go virtually unchecked until the late 1960s, when a series of potential environmental disasters created a sense of real urgency.

Initial Environmental Problems Associated with Florida's Growth, 1945–1970

Muck Fires

Before the construction of drainage canals and laterals during the period 1910–1930, muck fires occasionally took place during the dry season just before the onset of winter. These fires, often started by lightning, would sweep through the sawgrass, and in some cases the peat muck from which the sawgrass grows would spontaneously catch fire. However, as long as the water table remained at the surface, these fires were rare and usually burned themselves out when they reached overflow lands. However, after drainage started, the water table subsided and these fires became the rule, not the exception. The muck fires raged day and night until brought under control by heavy rainstorms. These fires affected not only the Everglades Agricultural Area south of Lake Okeechobee, but they also prevented water from flowing south, which historically had prevented the lower glades from drought.[55] By the mid-1940s the muck fires were so bad that they would create dense clouds of smoke that frequently drifted over Ft. Lauderdale, Miami, West Palm Beach, and other coastal cities. In 1970 one Florida official stated that there were times when the smoke over Miami "literally shut out the light of day," just like the Kuwait oil fires did in Kuwait City after the 1991 Desert Storm war.[56]

Soil Subsistence

The muck fires not only burned off the sawgrass, but they also destroyed the muck itself. In addition to fire, drainage helped to destroy the muck because organic soils are susceptible to biochemical oxidation after they have been drained. By 1940, the surface of muck soils near Lake Okeechobee had a depth of only seven or eight feet, compared to 12–14 feet in 1912. In some places where muck had never been deeper than one or two feet it had altogether disappeared.[57] As drainage continued, by the late 1940s one foot of muck was being lost every 10 years.[58] It was widely feared by Glades farmers that unless something was done, the loss of soil would mean that by the year 2000 most of them would be out of business.[59] Also, the loss of muck and soil from the Everglades due to drainage did irreversible damage to the region's bird and animal life.

Heavy Flooding

Rainfall in Miami averages 58 inches annually, but in 1947 there was a total of 102 inches, almost twice the average. In Dade County, communities including Hialeah, Miami Springs, and Northwest Metropolitan Miami were inundated. Ft. Lauderdale was also badly hit, with 30% — including all of the downtown — flooded. Property damage was estimated at $60 million, crops were destroyed, cattle drowned, and buildings severely damaged. After the flooding, the U.S. Army Corps of Engineers worked with local officials to create the Central and Southern Flood Control District, covering 18 counties with a total of 15,673 square miles. The project plan called for a levee paralleling the coastal ridge, and reserving 800,000 acres for agriculture, but designating 1.3 million acres for water conservation to maintain a better ecological balance. The project plan called for building new canals, pumping stations, and control dams. The project was virtually completed by the early 1960s and there were some startling successes. First of all, both the agricultural areas in the Glades and urban areas on the Atlantic Coast were protected from floods. Second, by conserving large areas of land in the Everglades in a natural state, these lands now were available for recreational use such as airboating and hunting. Also, by accident, with the initial construction of the canals now joined by sawgrass marshes and the entire area stabilized and relatively free from flooding and drastic changes in the water table, a perfect environment was created for large-mouth bass fishing. But there were some serious failures in this project's development: (1) the lack of a cost sharing arrangement that would be equitable for agricultural and urban interests — the urban areas wound up with higher taxes and minimal benefits; (2) the failure to grant a higher priority to storage and delivery of water necessary to meet anticipated demands for the proposed Everglades National Park as opposed to immediate needs of agricultural interests; and (3) lack of policies to keep land reclamation from increasing water demands beyond the land's capacity and encouraging development beyond the land's environmental tolerance.[60]

By 1970, these environmental problems were joined by other problems common to all growth areas across the United States: pollution of air and water, overcrowded roads, flooding, inadequate schools, parks, and public facilities, power shortages, and even difficulty in getting dial tones on business and residential telephones. The decline in the quality of Florida life, especially in the urban areas, was all too clear to be ignored. This decline is best captured by Ben Funk, an Associated Press writer who lived in Miami from 1951 to 1971.

> It was January 1951. Our destination was Miami — that special haven of Northerners seeking respites from the cold and grime back home. We hoped to remain all our lives....
>
> We were struck by the sight of seaside towns spread out in pastel colors over white sands. And in Miami, we found everything we sought. The waters were pure, the air free of contamination, the fishing great,

the life jolly and unhurried. The ten mile drive home from the office was a fifteen-minute breeze over roads and streets shaded by spreading banyan trees and flowering poincianas. Key Biscayne was a wild coconut plantation inhabited by raccoons and a small colony of humans who had settled on the island to get away from it all.... We could stroll through hardwood hammocks teeming with wildlife, swim in any body of water, walk a lonely beach, and watch the moon rise.

In those days, the tourist "season" was December to March. The rest of the year, most of the hotels closed and the town belonged to the "natives." It was one great place to live. But the drumbeaters were busy. Resort owners and land promoters, cities, airlines, counties, and the state were spending huge sums of money to lure people to this unique corner of the continent. They were coming in rapidly growing numbers.... More and more people meant more garbage, more trash, more sewage, more hotels, more automobiles. Human wastes poured in the ocean. Fumes from automobile exhausts and jet planes fouled the once pure air. Hammocks were flattened by the bulldozers. Dredges mangled the shorelines. Mile after mile, hotels and motels marched up the beaches, gaudy monuments to the tourist dollar hiding the ocean behind a concrete wall....

And one day we took a good look around. Suddenly we knew that we had kissed the good life goodbye.[61]

The Inability of Florida State Government to Deal with Problems of Growth

Between 1950 and 1970, Florida grew by just over four million residents. The percentage of "urban" residents increased from 56.5% in 1950 to 81.7% in 1970. However, state government was slow to deal with the problems of growth. There were two general reasons for this. One was the huge distances between Florida cities, making it difficult for individuals with common interests to join forces and maintain communication. Second, the Florida legislature was, up until the late 1960s, controlled by a tightly knit network of elected officials from North Florida known as the "Pork Chop Gang."

In 1949, Professor V.O. Key noted that Florida's unique geography prevented the emergence of strong machines that could make statewide decisions about matters like growth management compared to those that existed in places like Virginia, Georgia, and Tennessee. Stated Key:

> Florida's huge area and its peculiar configuration obstruct the formation of state wide political organizations. While great weight should not be assigned to this factor, the physical inconvenience of assembling persons from all parts of the state for factional collaboration needs to be pointed out.[62]

The roadway distance from Pensacola to Key West is greater than from Pensacola to Chicago. Jacksonville is further from Miami than Springfield, Illinois is from Columbus, Ohio. Even relatively close cities such as Tampa and Miami are 220 miles apart; the same distance as Boston and New York City. In the days before the availability of cheap long distance telephoning, fax machines, and safe, four lane, divided and controlled access highways, it was indeed difficult to assemble a statewide body to develop consensus on issues.

A better reason was the stranglehold over all of Florida by the "Pork Chop Gang" — a group of legislators aptly named because when appropriations were handed out they got the best cuts for their home constituents. They were a group of North Florida rural oriented politicians who controlled the Florida House and Senate. Even though population in Florida had shifted drastically from North Florida and the Panhandle to the peninsula between 1900 and 1960, the "county unit" rule created a situation where rural legislators dominated the more populous South. For instance, receipts from racetracks were evenly divided between all 67 counties even though by 1960 Dade had 935,000 residents and Glades, the smallest, had only 4,000. As late as 1965, one senator from Dade County represented 311,000 people, whereas another senator from the rural North represented only 17,000. The Pork Choppers — most of whom were from rural areas — were heavily influenced by agricultural, extractive, and development oriented interests, including those in lumber, pulp and paper, cattle, citrus, and phosphate. Ed Ball, head of the DuPont interests in Florida, which included a string of banks, the Florida East Coast Railway, and the St. Joe Paper Company, was a powerful behind-the-scenes player in the Pork Chop scenario. Potentially competing voices such as organized labor were very weak, and conservation interests were local groups interested only in local issues.[63]

However, things were to change in the late 1960s. First of all, in 1962 the Supreme Court, in Baker v. Carr, ruled that legislature districts in states must represent equal populations — the "one man one vote rule." In 1966, Florida voters, mainly from the South — Tampa, St. Petersburg, Orlando, and Miami-Ft. Lauderdale — tired of the antics of the Pork Choppers, elected the first Republican governor since reconstruction: Claude Kirk. Kirk pushed through a new Constitution that shrank some 200 state agencies to 25, permitted the governor to succeed himself, and strengthened the governor's office by placing two key departments under his direction: administration that contained the budget division and the new Bureaus of Planning and Transportation. In 1967, Baker v. Carr was finally implemented, giving the "urban" legislators power in numbers and breaking the Pork Chopper's stranglehold. Also in 1967, the legislature gave all counties in Florida the power to prepare zoning ordinances, whereas beforehand over one-half had no growth control mechanisms whatsoever. In 1967, the reconstituted Legislature enacted the "Sunshine Law," which mandated that all meetings of state and local government bodies at which business is conducted are to be public. This law was interpreted by the attorney general and the courts as applicable to any conversations between elected officials in which public business is a subject of discussion.

So, with a new legislature in place representing urban interests, state government streamlined, a governor capable of seeing his/her initiatives implemented over an eight-year period and rudimentary land use controls in place, Florida was now primed for a revolution in growth management practices.

Notes

1. William Bartram, *Travels of William Bartram,* Mark Van Doren, Ed. (New York) Dover Publication, Inc., 1955, pp. 100–211.
2. Luther Carter, *The Florida Experience, Land and Water Policy in a Growth State,* (Baltimore) Johns Hopkins Press, 1976, p. 4.
3. V.O. Key, *Southern Politics in State and Nation,* (New York) Alfred A. Knopf, 1950, p. 82.
4. Charlton W. Tebeau, *A History of Florida,* (Miami, Florida) University of Miami Press, 1971, pp. 110–178.
5. Ian McHarg, *Design with Nature,* (New York) Natural History Books, 1969.
6. Carter, op. cit. Note 2, pp. 57–58.
7. Ibid., pp. 58–59.
8. Ibid., p. 59.

9. See John DeGrove, *Land Growth and Politics,* (Chicago) Planners Press, American Planning Association, 1984, pp. 177–290; and Robert G. Healy, *Land Use and the States,* (Baltimore) Johns Hopkins Press, 1976, pp. 64–102.
10. Carter, op. Note cit. 2, p. 62.
11. Carter, op. cit. Note 2, p. 62.
12. Buckingham Smith, *Report on the Everglades,* S. Doc. 242 30 Congress (1848) in S. Doc. 8962 Cong. 1st session (1911), pp. 46–54.
13. Ibid., pp. 56–58.
14. Carter, op. cit. Note 2, p. 62.
15. Ibid., pp. 62–63.
16. It is important to note that of all statements regarding Everglades drainage mentioned in the Smith Report of 1848, only one, that by Stephen R. Mallory of Key West, the customs inspector who later became a U.S. senator and then Secretary of the Confederate Navy, questioned the prevailing thought. Mallory felt that considerable acreage in the Everglades was at sea level and could not be drained. See Carter, op. cit. Note 2, p. 63.
17. Carter, op. cit. Note 2, pp. 63–64.
18. Ibid., pp. 63–64.
19. Ibid., p. 64.
20. Ibid., pp. 64–65.
21. Tebeau, op. cit. Note 4, p. 274.
22. Carter, op. cit. Note 2, pp. 65–66.
23. According to the U.S. Census Bureau terminology operative in 1900, "urban" was defined as "all persons living in urbanized areas and in places of 2,500 or more persons outside urbanized areas."
24. Tebeau, op. cit. Note 4, p. 15.
25. J.E. Dovell, *A History of the Everglades of Florida,* 1947, unpublished doctoral dissertation: University of North Carolina.
26. Tebeau, op. cit. Note 4, pp. 18–19.
27. Ibid., p. 19.
28. Carter, op. cit. Note 2, p. 66.
29. Samuel Proctor, *Napoleon Bonaparte Broward,* (Gainesville) University of Florida Press, 1950, p. 190.
30. Tebeau, op. cit. Note 4, p. 20.
31. Ibid., p. 34.
32. Carter, op. cit. Note 2, p. 69.
33. Tebeau, op. cit. Note 4, p. 35.
34. Alfred J. Hanna and Kathryn A. Hanna, *Lake Okeechobee* (New York) Bobbs, Merrill and Company, 1948, p. 139.
35. Tebeau, op. cit. Note 4, p. 36.

36. Hanna and Hanna, op. cit. Note 34, p. 146.
37. Carter, op. cit. Note 2, p. 71.
38. In 1991 the CBS show "60 Minutes" ran an episode entitled "This House is for Sale." The episode depicted land sales techniques of the General Development Corporation. This corporation sold over 200,000 acres in Southwest Florida during the 1960s and 1970s to people sight unseen in a manner almost identical to Bolles group in 1908–1912, except this time the prices were much more expensive, even when inflation was factored in. General Development filed for bankruptcy protection in 1990, one step ahead of thousands of creditors.
39. Tebeau, op. cit. Note 4, p. 38.
40. Carter, op. cit. Note 2, p. 74.
41. Ibid., p. 74.
42. Polly Redford, *Billion Dollar Sandbar,* (New York) E.P. Dutton, 1970, pp. 71–73.
43. Ibid.
44. Carter, op. cit. Note 2, p. 76.
45. Ibid., p. 76.
46. Tebeau, op. cit. Note 4, p. 58.
47. Carl Grismer, *Tampa, A History of the City of Tampa and the Tampa Bay Region of Florida,* (St. Petersburg) D.B. McKay, 1950, pp. 88–116.
48. Tebeau, op. cit. Note 4, pp. 58–59.
49. Ibid., p. 63.
50. Carl Donovan, *A Decade of Federal Expenditures in Florida,* USGPO, Economic Leaflets I, January 1942, pp. 2–4.
51. Tebeau, op. cit. Note 4, p. 78.
52. Ibid., p. 79.
53. Four Fruitful Years: Florida Highways XIII (December 1944).
54. Tebeau, op. cit. Note 4, p. 135.
55. Carter, op. cit. Note 2, p. 85.
56. Spessard C. Holland testimony before the Subcommittee on Flood Control, Rivers and Harbors, Committee on Public Works, Hearing on Central and Southern Florida Flood Control Project, April 8, 1970, p. 51.
57. J.H. Davis, "The Peat Deposits of Florida," Florida Geographical Survey Bulletin, 301-242 1946.
58. John C. Stevens, "Peat and Muck Drainage Problems," *Journal of Irrigation and Drainage Division,* Proceedings of the American Society of Civil Engineers Annual Meeting (June 1969), p. 297.
59. Carter, op. cit. Note 2, p. 86.
60. Ibid., pp. 92–96.

61. Ben Funk, "Miami: Kissing the Good Life Goodbye," *Tallahassee Democrat,* March 21, 1971, in Luther Carter, op. cit. Note 2, pp. 146–147.
62. V.O. Key, Jr., *Southern Politics in State and Nation,* (New York) Alfred A. Knopf, 1950, p. 83.
63. Carter, op. cit. Note 2, pp. 44–45.

3 Toward Managed Growth: 1971–1991

1971–1975: From Hands Off to Mandated Planning: Florida's 180 Degree Turnaround

The National Mood

By the late 1960s social activists began to turn their attention toward environmental issues. The environmental movement began in earnest in the 1960s as a revival of the old conservation movements that created the Sierra Club and initiated our system of National Parks in the 19th century. A mounting body of scientific knowledge was, by the 1960s, linking pollution of the air and water to illness. Rachel Carson's *Silent Spring*[1] galvanized natural scientists, such as biologists and chemists, as she described the loss of natural habitats for birds and wildlife as a result of wanton pollution. Ian McHarg, in *Design with Nature,* invigorated professionals in fields such as architecture, landscape architecture, and city planning with his notions that now, instead of continuing the destruction of our environment, physical design and the built environment should be in harmony with nature.[2]

In 1969, Senator Henry M. ("Scoop") Jackson of Washington, a state with a strong activist environmental lobby, spearheaded the drive for the National Environmental Policy Act, which would require environmental impact studies for all major projects built with federal support or requiring federal permits. After a bitter fight with lobbyists for the National Association of Manufacturers, the U.S. Chamber of Commerce, and major mining, timber, and agribusiness groups, this act was passed and signed into law by President Nixon in December 1969. This legislation put the federal government on the side of the environmentalists and set the stage for a remarkable turn of events in Florida's land development policy orientation.

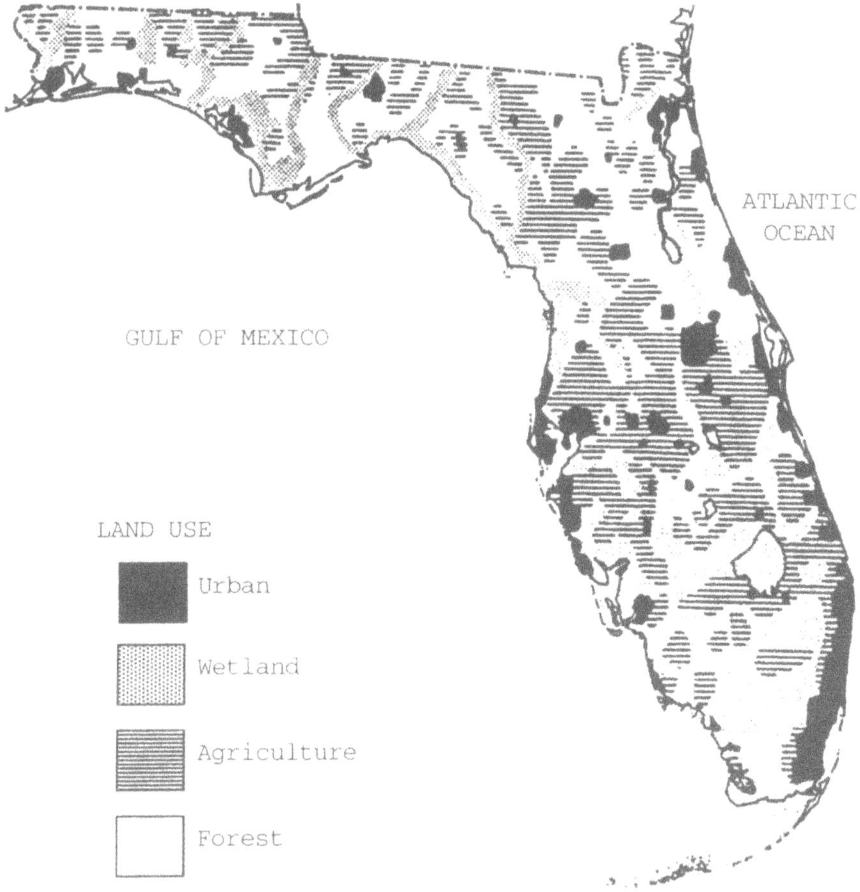

Figure 1. Land use in Florida, 1970. Source: Luther Carter: The Florida Experience, *Johns Hopkins Press. Reprinted with permission.*

Environmental Victories in Florida—1968–1971[3]

The changing national mood toward environmental protection came at a time when local environmental groups began challenging vested interests in Florida, including railroads, agribusiness, tourism, and industrial elites. Three cases illustrate this phenomenon: the Florida Barge Canal, the Miami Jetport, and the fight to preserve the water supply for Everglades National Park.

The concept of constructing a barge or ship canal across North Florida using the St. Johns and Oklawaha rivers to connect the Atlantic with the

Gulf went back as far as the time when Florida was acquired from Spain as a territory. During the Depression, President Franklin D. Roosevelt initiated acquisition of the right-of-way and appropriated $5 million to start construction. The money was used up in one year just to clear a headquarters site known as Camp Roosevelt and initiate infrastructure for a few bridges. Work did not begin again on the canal until 1964 and, between that time and 1969, considerable progress was made. The Rodman Dam was built in the lower Oklawaha Valley, and at its rear a 306-ton crawler crusher knocked down 6,500 acres of swamp hardwood forest into the mud and prepared a reservoir to be filled. By mid-1969 this reservoir, known as Rodman Pool, was completed, but logs would occasionally rise to the surface, making this "lake" unsuitable for boating and fishing. Also, three of the canal's five locks were completed by mid-1969 and the excavation of the canal itself was partially finished. Over $50 million had been expended, and the battle to save the Oklawaha River and its natural ecosystem appeared to be lost.

However, the environmentalists banded together as the Florida Defenders of the Environment, established a liaison with the national Environmental Defense Fund, and sued the U.S. Army Corps of Engineers, demanding that an environmental impact statement be prepared as per the requirements of the National Environmental Policy Act that had just been passed. The resultant publicity created a groundswell of opposition to continuing the project, the environmental impact statement did not support construction and the President's own Council for Environmental Quality recommended that it be terminated and, on January 19, 1971, President Nixon stated, "the Council on Environmental Quality has recommended to me that the project be halted and I have accepted its advice. The canal...would destroy a uniquely beautiful semitropical stream." The project was indeed halted; no more work was done, and 20 years later, in 1991, a state grant was given to the University of Florida's College of Architecture to design a regional park with hiking trails along the old canal path.

By the mid-1960s, Miami's airport was reaching its capacity and the Dade County Port Authority began to look for a new "jetport" site. By 1967, the Authority had selected a 40 square mile site in the eastern portion of the Big Cypress Swamp about 45 miles west of Miami and six miles from Everglades National Park. The original plan called for as many as six long runways, two of which would be 30,000 feet, or almost six miles. Construction began in 1968 with the completion of the first step, a single training runway finished in early 1969. The chairman of the Central and Southern Flood Control District, Robert W. Padrick, objected to a road connecting the airport with Miami, as it would cross the Flood Control District and create environmental problems. He called upon environmental groups nationally and locally to assist him in opposing this airport. With the national environmentalist movement in full swing, a coalition developed that within months included 21 environmental groups and two labor unions, which would then pressure Governor Claude Kirk into opposing the airport site. The environmental impact statement required by the new National Environmental Policy Act of 1969 concluded that the site would eventually, as a result of adjacent urban development, have a negative impact on the Big Cypress Swamp and Everglades National Park. The Port Authority then abandoned the Big Cypress site and looked for an alternative in several locations around Dade, Broward, and Palm Beach Counties. When no alternative could be found that was acceptable to local residents, environmentalists, and the Port Authority itself, plans for a new airport were abandoned. Miami International was expanded to capacity between 1970 and 1980 and new enplanement demand was met by shifting traffic to airports in Ft. Lauderdale and West Palm Beach, both of which were expanded and modernized during the period of 1980–1990.

Everglades National Park was established in 1958 after a long struggle between conservationists, developers, and state government that began in the mid-1920s. The Florida legislature finally provided $2 million for the purchase, totaling 1.3 million acres, in 1947. This park receives most of its water by natural rainfall, but the flow of water over land from north of the Tamiami Trail into the park is important in maintaining its natural ecology. The Central and Southern Flood Control District established by the U.S. Army Corps of Engineers, working with local government just after the heavy floods of 1947, had a policy to release water for the park only when storage in water conservation areas was at maximum or near maximum capacity. This policy protected urban and agricultural water needs at the expense of the park. In 1970 Congress solved the problem by authorizing funds for the enlargement of the Miami Canal, and upon completion of that project the park would receive an amount of water to satisfy its ecological needs without infringing on the needs for urban and agricultural use. This legislation was another victory for the newly organized and now energized Florida environmentalists.

The Election of Reubin Askew

Since Reconstruction, Florida had been led by various factions of traditionalists: Bourbons, Elitists, and the Pork Choppers. As V.O. Key noted, there was never a strong single leader in the state due to geography and the fact that governors were limited to a single term. This created a situation where land development decisions were made by obscure boards such as the Internal Improvement Fund and later coalitions within the Florida Cabinet, a group that consists of the secretary of state, the attorney general, the comptroller, the state treasurer, the superintendent of public instruction, and the commissioner of agriculture. These officials were elected every four years at the same time as the governor, *but, unlike the governor, they could be reelected indefinitely.*[4] These individuals purposely maintained a low profile and were singularly responsible for the land giveaways — first to the railroads between 1880 and 1905, and then bargain basement sales to people like Disston in 1881, Bolles in 1908, Fisher in 1917, and countless other speculators up until the 1960s.

By the mid-1960s, Florida's 6.5 million residents, 80% of whom lived in urban areas, were ready for reform. Almost half were migrants from the Middle West, the northeast megalopolises, and Canada, and they were used to a more progressive form of government. The trend had already been established in the 1950s when Florida had given a majority of its votes to General Dwight D. Eisenhower in the 1952 and 1956 presidential elections, thus breaking with the old Democrat-controlled "solid South." In 1966, Claude Kirk became the first Republican governor since Reconstruction. Between 1966 and 1970, as stated before, several earthshaking events took place in Florida politics, including implementation of Baker v. Carr, a new constitution, permitting, among other things, the right of the governor to succeed himself, and government in the sunshine, all of which destroyed the old Pork Chop Gang's domination of state politics. All of these events, coupled with national movements supporting environmental protection and the rights of women and minority groups, made for a climate of change.

In 1970, two obscure Florida legislators, Reubin Askew and Lawton Chiles, made bids for Governor and U.S. senator, respectively. Both were state senators—Askew from Pensacola and Chiles from Lakeland. Chiles pledged to walk every county in the state to gain a personal understanding of Florida's problems. His campaign drew national publicity and he was soon dubbed "Walkin' Lawton." Askew was a bit less flamboyant, but he quickly staked out a bold liberal program for Florida, pledging to install a state corporate income tax and comply with the then recent U.S. District Court decisions to abolish public school segregation with the use of bussing if necessary. Askew — young, well educated, and articulate — caught on not only with the younger voters, ex-Northerners, retirees, and blacks, but also with a sizable number of rural "crackers" who quietly resented being ripped off by special interests, especially when it came to issues such as taxation and provision of health and social services. A newspaper article on October 17, 1970, quoted an elderly white woman in a low income area in Palatka as saying, "It just won't set well with me if these boys lose," referring to Askew and Chiles.[5] This lady was not disappointed, as Askew easily defeated incumbent governor Claude Kirk and Chiles was elected over an incumbent Republican senator.

Once in office, Askew initiated implementation of his campaign promises and won instant national attention. The May 31, 1971, issue of *Time* magazine featured Askew along with four other newly elected Southern governors — John West of South Carolina, Linwood Holton of Virginia, Dale Bumpers of Arkansas, and Jimmy Carter of Georgia. Calling these five men "the new breed of Southern governors," *Time* wrote that they have departed from the old practices of racism and provincialism and favor economic development balanced with environmental sensitivity and new directions in race relations. Askew was quoted as saying "last fall's elections represent a departure from the custom wherein the person who took the hard racial line always won."[6]

Every movement needs a leader—one individuals and groups may rally around, one who can articulate emotions that people feel but somehow cannot bear to bring to the surface—someone they can trust to do what is right rather than what is politically correct in the short run. First, Samuel Gompers and then John L. Lewis and the Reuther brothers provided this leadership for organized labor. The Reverend Dr. Martin Luther King, Jr., provided this leadership for the civil rights movement, and Reubin

Askew provided this quality of leadership for the revolution in Florida's land development policy.

The Land and Water Management Acts of 1972

In 1970 and 1971 Florida suffered a massive drought. The extreme drought brought muck fires in the Everglades and raised the specter of saltwater intrusion into the Biscayne Aquifer once again. This water crisis, pronounced by observers as the worst ever, prompted a statewide conference convened by Governor Askew in August 1971 in Miami Beach. It was attended by over 150 participants, including developers, state and local government officials, federal agency representatives, and environmentalists. Askew, a man of vision, saw the conference as a "lens" by which one could closely examine the problems and propose solutions for the much broader issues relating to growth management. In addressing those who offered only a cautious approach, one that would not negatively affect the agendas of developers and agribusiness interests, Askew replied, "It is time we stopped viewing our environment through prisms of profit, politics, geography, or local and personal pride."[7] He warned that "a failure to find appropriate solutions to the effective management of growth would be disastrous to our economy as well as to our environment."[8] The conference responded to Governor Askew's theme by drafting a strongly worded set of findings and policy recommendations. The report stated that "an enforceable comprehensive land and water use plan...must be designed to limit increases in population...to a level that will ensure a quality environment."[9]

The 1972 legislative session continued the conference's work by passing four pieces of legislation, all of which involved compromises. They were as follows:

1. The Environmental Land and Water Management Act

This act was based on a concept of joint state-local regulation of (1) areas of critical environmental concern and (2) Developments of Regional Impact (DRI). Areas of critical concern could only be designated for (a) areas containing or having a significant impact upon environmental, historical, natural, or archaeological resources of statewide importance, (b) areas significantly affected by or having a significant effect upon an existing or proposed major public facility or other area of major public investment, and (c) a proposed area of major development potential which may include a proposed site of a new community stated in a state land development plan.[10] In order to get the critical areas portion passed, only 5% of the state's land could be designated "critical" at any one time.

The development of regional impact provision was designed to deal with large projects, such as major housing developments, airports, power plants, large shopping centers, and office parks that "because of [their] character, magnitude, and location would have a substantial effect upon the health, safety or welfare of citizens of more than one county." Figure 2 shows the original decision making structure for regulation of areas of critical concern and Developments of Regional Impact. As one can see, in areas of critical concern, while the local government initiates the process, state government plays a major role, as it can review and modify local governments' regulatory actions. On the other hand, Developments of Regional Impact process rests primarily with the initiative and activity of developers and local government, as the regional planning agency simply prepares the regional impact statements, which are not binding on local government. Therefore, if a local government is developer oriented, it would be difficult to maintain strong regulation. The local government is charged with responsibility for deciding whether (a) the development unreasonably interferes with the achievement of the objectives of an adopted state land development plan, (b) the development is consistent with the local land development regulations, and (c) the development is consistent with the report and recommendations of the regional planning agency.[11] The ultimate responsibility for DRI land use decisions rests with local governments, and local government can ignore the regional agency's recommendations. Even though the act provided an appeal to the cabinet for planners and environmentalists, appeals to this board dominated by "good ole' boy" types who tend to favor development and its resultant profits was like allowing the foxes to guard the chicken coop.

2. The Water Resources Act (See Figure 3)

This act established five regional water management boards governed by the Department of Natural Resources and the Florida Cabinet to regu-

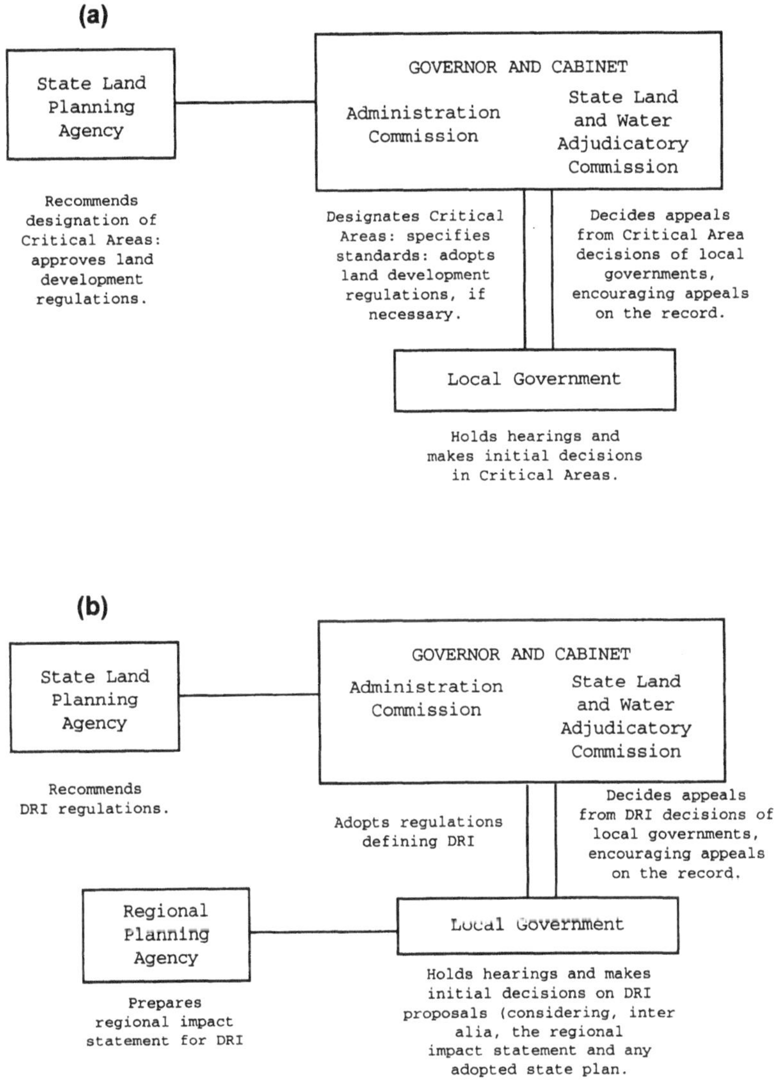

Figure 2. *(a) Decision-making structure for regulation of critical areas under the 1972 Land and Water Management Act. (b) Structure for regulation of developments of regional impact. Source: Gilbert L. Finnell, Jr., "Saving Paradise: The Florida Environmental Land and Water Management Act of 1972,"* Urban Law Annual *(St. Louis, Missouri: Washington University, 1973). Reprinted with permission.*

late water use by local drainage districts, county water management boards and, in times of emergency, the consumption of water. With the districts covering the entire state, Luther Carter felt that this legislation would finally bring an end to "the near anarchy that has characterized land drainage and other water management activities in many parts of Florida."[12]

3. *The Land Conservation Act*

This act authorized the issuance of $200 million in state bonds to purchase "environmentally endangered lands" and $40 million in bonds for outdoor recreation areas. In the November 1972 general election, Florida voters passed these bond issues by a majority of just over 70%. However, as Robert Healy points out, a precedent was set whereby the state was obligated to purchase environmentally sensitive land rather than just regulate its use.[13]

4. *The Florida Comprehensive Planning Act*

This act permitted development and adoption of the State Plan — a policy document that would provide long-range direction for the "orderly social, economic, and physical growth of the state by articulating goals, objectives, and policies." The Plan was to become official once approved by the governor, who was designated under the Act as the chief state planning officer, and by the legislature. The Department of Administration and its new planning division were to coordinate the various state agency budgets with the Plan. The Planning Act did not provide for the state to override local decisions when they conflicted with the State Plan. However, this act was never fully funded and the adopted State Plan was not completed and finally adopted until 1985.

The Local Government Comprehensive Planning Act of 1975 (LGCPA)

In 1972, the legislature, realizing that so much more needed to be done, set up a group of its membership known as the Environmental Land Management Study (ELMS) Committee as part of the Land and

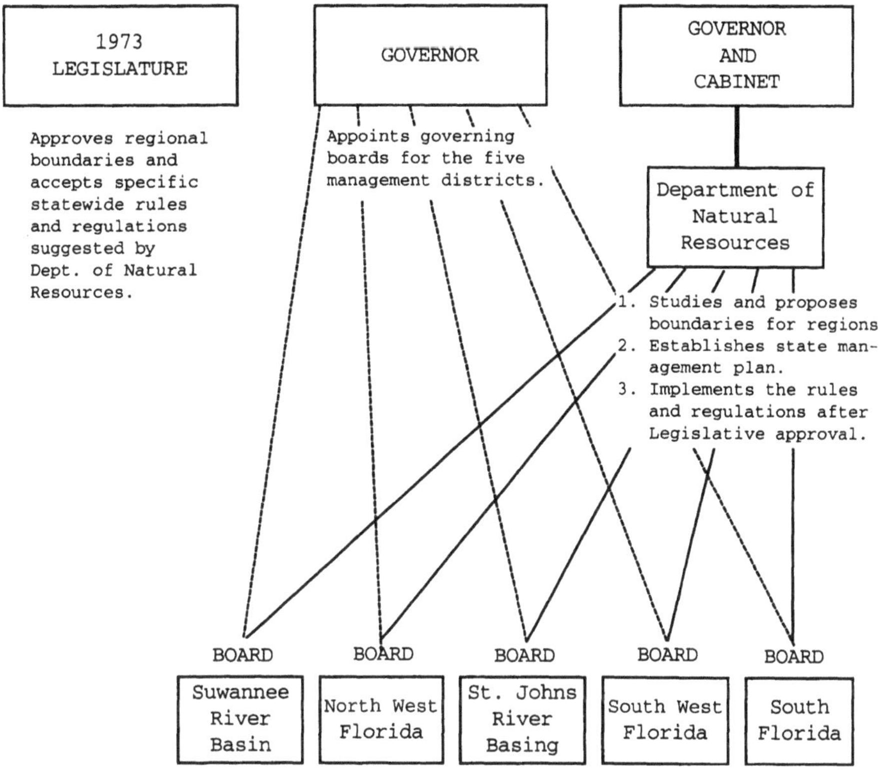

Figure 3. Decision-making structure established under the Water Resources Act of 1972. Source: ENFO Newsletter, *April 1972, Environmental Information Center, Winter Park, Florida. Reprinted with permission.*

Water Management Acts package. This committee was quick to recognize that in order to enforce the DRI criteria of "the development must be consistent with local land development regulations," each unit of local government must have an up-to-date, enforceable comprehensive plan. In 1972, only one-half of Florida's counties had development regulations of any kind. Few had comprehensive plans. The local comprehensive plan would close a major loophole in the DRI process. DRI regulations state that if no recommendations are issued within 90 days by local government the developer may proceed anyway. With a majority of local governments having neither zoning, subdivision regulations, or an approved plan to *base* recommendations on, a developer had an easy way out. The

ELMS group in fact drafted a Local Government Comprehensive Planning Bill in 1974. Previous laws had *"permitted"* local governments to prepare comprehensive plans, but now this law would *require* all units of government to adopt one and, once doing so, no zoning ordinance could be in conflict with the plan and no development that was in conflict with the plan could be approved, *even if it met other Development of Regional Impact criteria.* Monies were to be provided to local governments to aid in plan preparation. Another recommendation of this group was that local impact fees be levied on developers to pay for infrastructure needs generated by their projects so that "growth would pay for growth" and that growth should not outpace or exceed the "carrying capacity" of natural and fabricated systems. This resolution would, by implication, call for state government to assume several new responsibilities, including getting policies for energy conservation, better distribution of residential, industrial, and tourism development within the state, preservation of prime farmland, and capacity building on the part of local government so they might assume primary responsibility for growth management.

These recommendations proved to be tough pills to swallow. In 1974, less than half a dozen cities and counties with populations over 10,000 had comprehensive plans prepared after 1960. The LGCPA would require 461 cities and counties to prepare plans in just three years. While 26 graduate programs in urban and regional planning had been established across the United States by 1960, the first planning program to be established in this state was at Florida State University in 1965, and by 1974 this program was graduating 15 to 20 planners per year. One must realize, though, that it takes anywhere from three to five years of post-graduation work experience to become a full-fledged professional planner. The University of Florida was scheduled to initiate a program in 1977 and the University of Miami began a small program in 1973, graduating five to ten students per year by 1978. Even if adequate funds were to be provided to prepare these plans, where were the planners going to come from, especially when it would take anywhere from six months to a year for a non-Floridian to become familiar with local environments and state/local land use history and present regulations? Most local planning staffs, even those in large cities such as Tampa, had no experience in preparing long-range plans, and instead had routinely confined their duties to reviewing zone change proposals, subdivision plans, and proposals for public acquisitions. In many instances, environmentalists blasted these planners as "expediters for the developers."[14]

However, the real opposition to the ELMS report came from development interests who realized that mandatory planning tied to zoning and the Development of Regional Impact criteria plus the imposition of impact fees would be the final nail in the coffin of "business as usual." The LGCPA was not passed by the 1974 Legislature and, as Reubin Askew was up for reelection that year, he chose not to overly pressure legislators to vote for the ELMS recommendations then. However, once reelected by over 60% of Florida's voters, Askew used the weight of his office to push through the Local Government Comprehensive Planning Act in the 1975 legislative session. The major difference between the 1974 act, which was not passed, and the 1975 version, was that in the case of the latter the state provided no funding for either plan preparation or state review. The provisions of this act were:

1. All municipalities and counties must prepare and adopt a comprehensive plan by July 1979. Upon showing of cause, the governmental jurisdiction could apply for a one-year extension.
2. Each unit of local government must, by July 1, 1976, designate a "local planning agency" to prepare its plan. In most cases, the LPA would be the county government. This provision allows small cities to avoid the expense and possible duplication of services in plan preparation.
3. Each plan must include the following elements:
   a. Future land use.
   b. Traffic circulation.
   c. Sanitary sewer, solid waste drainage, and potable water.
   d. Conservation.
   e. Recreation and open space.
   f. Housing.
   g. Intergovernmental coordination.
   h. Utility.
   i. Coastal zone protection (where applicable).
   j. Mass transit (communities over 50,000).
   k. Port, aviation, and related facilities (communities over 50,000).
4. Optional elements could include (a) Non-automotive vehicular and pedestrian traffic; (b) Off-street parking facilities; (c) Public services and facilities; (d) Community design; (e) General area redevelopment;

(f) Safety; (g) Historical and scenic preservation; and (h) Other, as determined locally.

5. As far as citizen participation was concerned, the only requirement in this area was for two public hearings — one on findings of existing conditions, including needs, and the second on the plan itself.
6. Once adopted, all land development regulations, including zoning ordinances, subdivision regulations, and building codes, must be consistent with the plan.[15]

Dr. Ernest R. Bartley noted in 1975, just after the LGCPA was passed, that two cautions must be observed: (1) there was no way of predicting the *quality* of plans to come from this act — some governments will do only the required elements, and others will use the act to really come up with plans that meet their community's needs; and (2) with no funding by the state for preparation of these plans, local communities will bear the burden of plan preparation and, of course, unless they are extremely conscious of the public good, will spend as little as possible to comply with the letter of LGCPA.[16] Dr. Bartley's observations proved to indeed be prophetic, as subsequent events and the case studies will show in Chapters 4–7.

Implementation of the LGCPA, 1975–1990

By January 1980 over 300 of Florida's 461 units of local government had submitted plans to the Florida State Department of Community Affairs for review and comment.[17] Governor Bob Graham, elected for his first term in 1978, convened a group known as the Resource Management Task Force shortly upon his inauguration, and one of that group's assignments was to review the quality of plans prepared as a result of the 1975 LGCPA. In his final report prepared in 1980, that group concluded that LGCPA's objectives had not been achieved because (1) most local plans had extremely vague goals and policies, making implementation extremely difficult; and (2) many plans were frequently amended, making them virtually useless after two or more years from initial adoption.[18]

In 1982, Governor Bob Graham created the second Environmental Land Management Study Committee. Known as ELMS II, this group was given the charge of reviewing the Environmental Land and Water Management Act (Chapter 380 F.S.) and all related growth management programs. ELMS II was also given the responsibility for preparing a new guide for growth and development in Florida for the 1980s and beyond. This committee made its final report in 1984, concurring with the findings of the Governor's Resource Management Task Force and adding five other points: (1) the lack of strong state and regional planning, (2) lack of adequate funding for preparation and implementation of local plans, (3) lack of consistency between local plans and state/regional plans and/or policy guidelines, (4) lack of quality standards for local plans prepared under the LGCPA, and (5) absence of effective implementation measures for these plans.[19] The ELMS II Committee also recommended in their final report (1) that local plans be made consistent with state and regional plans, (2) that limited citizen standing be given to enforce the consistency requirement of the LGCPA, (3) a tightening of the plan amendment process, (4) a requirement for a land use map for all plans, and (5) setting up a growth management trust fund to provide monies for state and regional planning agencies and local government in the preparation and implementation of required plans.[20]

Governor Graham was handily reelected in 1982, and armed with recommendations from his Resource Management Task Force, the ELMS II Committee, and the consensus reached by a conference of over 120 top level representatives from state and local government, business, development interests, and citizen groups at Saddlebrook Resort in 1983,[21] was prepared to overhaul Florida's growth management program. In its 1984 session, the Florida legislature passed the State and Regional Planning Act.[22] This act required the Executive Office of the Governor to submit to the legislature a state plan consisting of goals and policies designed to give specific direction to state and regional agencies with responsibilities in growth management. The State Comprehensive Plan was adopted by the legislature and signed into law by Governor Graham in 1985. It addressed four major groups of issues: "Our People, Our Natural Environment, Our Communities, and Our Economy." In essence, the Florida Comprehensive Plan was not simply a physical policy plan, but one that addressed human service needs as well—something missing from most local plans prepared under LGCPA.[23]

In 1985, the legislature approved, and Governor Graham signed into law, the *Omnibus Growth Management Act of 1985*. The act required that (1) all units of local government prepare and adopt comprehensive plans, which must be consistent with regional policy plans adopted by regional planning councils, which in turn must be consistent with the State Comprehensive Plan, (2) all state agencies were required to prepare functional plans consistent with the state comprehensive plan, (3) all land development regulations adopted by local government must be consistent with the adopted comprehensive plan, and (4) a system of state review and approval of all local regional and state agency plans was provided for in what John M. DeGrove and Julian Juergensmeyer call "top down" consistency.[24]

The provision of the Growth Management Act relating to the preparation of local plans was entitled *The 1985 Local Government Comprehensive Planning and Land Development Regulation Act.* In addition to being consistent with the state comprehensive plan, the new local plans had to meet the following requirements, which differed from those in the 1975 LGCPA:[25]

1. All elements have specific measurable objectives to evaluate their effectiveness.
2. All plans would have to meet "minimum criteria" promulgated by the Department of Community Affairs. If the minimum criteria requirements were not met, local governments would be found in noncompliance, with penalties and sanctions imposed by the state.
3. A mandatory capital improvements element.
4. The future land use map must be included.
5. The plan, once adopted, could be amended only twice in one year.
6. Provision was made for limited citizen standing to challenge the plan.

In order to address the "minimum criteria" need that was designed to upgrade the quality of local plans, the Florida Department of Community Affairs on February 15, 1986, adopted R9j5, consisting of over 70 pages. R9j5 detailed the types of data, analyses, goals, objectives, and policies that were to be included in each plan element. The plan elements covered by R9j5 were future land use, traffic, circulation, mass transit, port aviation and related facilities, housing, sanitary sewer, solid waste, drainage, potable water, natural groundwater aquifer recharge conservation, recreation and open space, intergovernmental coordination, coastal management (where appropriate), and the new capital improvements element. There were also specific data requirements for consistency with comprehensive regional policy plans and the State Comprehensive Plan.

No sooner was R9j5 adopted than the critics came forth with varying observations, many of which were well founded. Legal scholars felt that R9j5, if pushed too vigorously by the state, would not hold up in court. Dr. Earl Starnes, of the University of Florida's Department of Urban and Regional Planning, noted that strengths included (1) the fact that standards and criteria were developed by planning professionals, (2) it was tailored specifically to fit Florida's needs, (3) it focused attention on special areas that would otherwise be overlooked by local government, and (4) it encouraged use of a standard format by all local governments. Dr. Starnes also noted some weaknesses, such as (1) costs to local government, at $22 million for compliance with R9j5 alone, (2) restriction of flexibility in local community planning, and (3) a limitation on planning quality, as local governments, especially those with limited funds, will spend money only to meet minimum criteria standards and not go beyond into new, different and innovative areas. As Dr. Starnes succinctly put it, "the minimum criteria rule may in effect become the maximum criteria."[26]

The legislature, in response to criteria of R9j5, passed what was termed the "Glitch Bill" in the 1986 legislative session. When R9j5 was submitted to the legislature for review and subsequent approval, it provided that compliance meant that a local plan met the requirements of R9j5 and Chapter 163 (Local Government Planning Acts) and was consistent with the State Comprehensive Plan and the appropriate regional policy plan.[27] But after listening to various interest groups, the legislature chose to straddle the fence by stating:

> For the purpose of determining whether local comprehensive plans are consistent with the State Comprehensive Plan and the appropriate regional policy plan, a local plan shall be consistent with such plans *if the local plan is "compatible with" and "furthers" such plans. The term compatible with means that the local plan is not in conflict with the State Comprehensive Plan or appropriate regional policy plan,* the term "furthers" means to take action in the direction of realizing goals or policies of the state or regional plan.[28]

As one can clearly see, there is a big difference between "consistent" and "compatible." The Glitch Bill also allowed for local governments to set their own standards for adequate infrastructure, within limits, and even if these standards did vary from state and regional levels of source, the local plan would still be deemed consistent because it was "compatible."[29] Other "Alice in Wonderland" provisions tacked on by the legislature stated that "while local policies must be based on data, those data are not subject to the compliance review process but DCA may use it to determine compliance" and DCA may not dictate a given methodology, but may determine if the one used was "professionally acceptable."[30]

Funds for the new Growth Management Act, especially the Local Government Comprehensive Planning and Land Development Act provisions, were in short supply. Comprehensive plan preparation, including compliance with R9j5, was estimated at about an average of $500,000 per unit of local government, which meant almost $230 million statewide, according to this author. The legislature did provide over $10 million in the 1985 and 1986 sessions, but the funds were quickly gobbled up by eager state and regional agencies and local governments. In November 1986, Bob Martinez was elected governor. With a mandate to reduce spending, Martinez led the fight to repeal the tax on services in 1987, and as this author noted in Chapter 1, although the Zwick Commission noted that $53 billion would be needed statewide by the year 2000 to pay for infrastructure improvements mandated by the 1985 Growth Management Act—$35 billion at the state level and $18 billion at the local level—repeal of the sources tax and the revenue enhancing measures caused shortfalls of $98 million in 1987–88 and $144.5 million in 1988–89.[31] Even with Martinez's defeat in 1990 — given the recession of the period of 1989–1992 — it is doubtful whether one can count on state government to help in providing funding even for plan preparation, to say nothing of needed monies to implement these plans.

By the end of 1990, 265 units of local government had submitted plans for review and approval to the Department of Community Affairs. Of these, 97 were approved without objection, 44 were approved after DCA objections were met without going to administrative hearings, 62 were approved as a result of the administrative hearing process, and 60 were still awaiting approval.[32] The impact of these plans on growth will not be known for years to come.

Summary and Conclusions: The Evolution of Land Development Policy in Florida, 1850–1990

We find that from the very beginning Florida was ripe for exploitation. Geographically isolated from the United States, the Peninsula was the last portion of our nation to be settled. Its far-flung reaches and individualistic/traditionalistic culture — which produced weak central political leadership or no leadership at all — combined to create a climate where development decisions were actually made by an unholy alliance of minor elected officials who could be reelected indefinitely (some members of the Florida Cabinet) and private entrepreneurs who carved up the state for their personal profit, while governors elected for one term only could only cut ribbons, looking like they were in charge but knowing otherwise. First, errors were made in surveying federal lands — meaning that sovereignty lands that should have been retained for public use and access were actually made available for speculation. Then, during the period of 1851 to the early 20th Century — disturbed only by the Civil War and Reconstruction — the Internal Improvement Board gave away millions of acres to railroad companies to entice them to build. The railroad companies held most of this land for speculation, and when the Progressive Era ushered in relatively liberal governors, including William Sherman Jennings (1901–1905) and Napoleon Bonaparte Broward (1905–1909) public policy shifted from railroad giveaways to blue light sales to hucksters such as Bolles, Fisher, and Davis, all promising to drain the land and then use it for agricultural or urban activities. Even when Florida became a truly "urban" state in the 1960s, government's role remained one of development promotion rather than regulation. While one would be quick to say this was the case all over America, particularly in the late 19th and early 20th Centuries, closer examination of the facts proves otherwise. In the period of 1890–1915 there was a strong conservation movement all over America — one that produced our National Parks and National Forests and men such as John Muir, founder of the Sierra Club, and Gifford Pinchot, who founded the National Forests program and urged a balance between conservation and economics development such as logging and mineral extraction. This movement almost completely bypassed Florida. In fact, while the first 15 National Parks were established between 1890 and 1897, Florida's Everglades National Park was not established until 1958, after a thirty-year battle between development interests and state government on one side and conservationists on the other.

Between 1966 and 1970 drastic political changes set the stage for corresponding change in land development policy. These included (1) Claude Kirk's election as governor — the first Republican to be elected to that office since Reconstruction, (2) implementation of Baker v. Carr, thus giving the urban legislators clout and eventually breaking control of the legislature by the Pork Choppers, (3) the 1968 constitution, which permitted the governor to succeed himself, and (4) "government in the sunshine." The rising environmental movement nationally and locally helped Florida environmentalists to win the battles in issues such as the Cross Florida Barge Council, the Miami Jetport, and the Everglades National Park, thus giving this group the confidence they needed in battling entrenched timber, agricultural, and urban development interests. Finally, the election of Reubin Askew as governor in 1970 set the stage for land development policy reform.

In four short years — 1971 to 1975 — sweeping legislation turned development policy from laissez-faire to mandated planning. *But did things really change?* The Land and Water Management Acts passed in 1972 and the Local Government Comprehensive Planning Act of 1975 must be examined more closely. Regarding the Land and Water Management Acts, if we examine the legislation that designated areas of critical concern, during the first year only 500,000 acres could be designated, and Luther Carter and Robert Healy tell us that the amount of land that could *eventually* be designated as being of critical concern was limited to 5% of the state's land, or about 1.7 million acres.[33] However, the greater loopholes were (1) developments previously authorized, such as Disney World, were exempt from this legislation, and (2) the final arbiter of decisions involving critical concern designation is the Florida Cabinet. With no guarantee that a pro-planning governor such as Askew would be on board forever, business interests who contribute heavily to campaigns of a majority of the cabinet members will have at least an initial advantage with respect to the Land Conservation Act of 1972. As I have stated earlier, Carter and Healy's concerns that a provision to purchase environmentally sensitive lands could encourage property owners to insist that a compensable development right adheres to all lands affected by protective orders issued under the act and in effect gives a right to monetary compensation rather than simple regulation.

With respect to Developments of Regional Impact (DRI), there are several flaws created by compromise in order to get the bill passed. First of all, the Regional Council's recommendations are not binding on local government. This is unfortunate, because only the very large cities and counties have adequate staff to accurately review DRI proposals, and the regional planning agencies, while somewhat influenced by local government officials, at least have staff capacity to adequately complete reviews with a fair to moderate degree of impartiality. Second, the DRI threshold varies considerably from place to place; a 100,000 square foot office complex could have a real impact in Glades County, with 4,000 people, but it would only be a drop in the bucket in Dade County. The way the law was initially written, developers could design their projects to drop just under the threshold and avoid a DRI review. Third, enforcement is a problem, as the state did not, at the time of the legislation's passage — nor does it now — have adequate staff to monitor projects. Fourth, the DRI process could not deal with the *cumulative* impact of a series of small projects, such as the multitude of shops that opened on Tampa's Dale Mabry Highway between 1980 and today and snarled traffic hopelessly. Fifth, projects approved before the legislation passed but not completed or even initiated were exempted from its requirements. Sixth, the Florida Cabinet calls the final shots: same note as that regarding areas for critical concern. Finally, while one of the evaluation criteria states that "the development must be consistent with local land development regulations," over one-half of all units of local government *had no land development regulations in 1972.* The Local Government Comprehensive Planning Act of 1975 sought to correct that loophole, but plans were not due until July 1979. That gave developers an additional four-year window to get their projects through the regulatory process without the inconvenience of having to be consistent with comprehensive plans.

The Local Government Comprehensive Planning Act of 1975 was replete with faults. Earlier I stated that Florida had an extreme shortage of trained planners familiar with the state, as the first Florida-based graduate level program in Urban and Regional Planning was not established until 1965, and even by 1975 few cities had experience with preparing long-range comprehensive plans because their focus was almost always on current planning: zoning, subdivision regulation, and public acquisition. Dr. Bartley stated in 1975, just after the act was passed, that there was no guarantee of uniform quality of these plans and the lack of funding would produce, at best, mediocre products. In the initial drafts of this legislation in 1974 there were fairly detailed requirements that would have improved the quality of plans: there were provisions for local enactment of impact fees, and there was a provision to provide up to $20 million to help fund their preparation. The 1974 act was defeated, and when it did pass in 1975 all three provisions were dropped — the second allegedly because of the recession that produced a revenue shortfall. But there were additional problems with LGCPA. First of all, there was no tie between planning and implementation. While zoning was to conform with the plan, what was to prevent planners from going to the zoning ordinance, rounding off the corners of various districts, and calling it a plan? Without standards as to what constitutes "good planning practice," this and other shenanigans were all too possible to discount. The fact that all state agencies could do was "comment" meant that they had no power to reject plans that did not meet obvious professional standards. Actually, the State Department of Community Affairs had until 1980 only one person who reviewed plan submissions. Second, there was no link between plan recommendations and the capital improvements to implement them. While the planning guide propagated by the State Department of Community Affairs spoke loosely to this need, the language was soft and C.I.P. was not part of the required elements. States the "Guide":

> Since most plans will be long range, a capital improvement program is likely to be necessary. Plan elements which include proposed capital improvements must contain fiscal requirements including estimated costs, priority ranking, and proposed funding sources.[34]

Still there was nothing to guarantee that if these capital improvements cannot be provided, the development could not proceed. Another problem was the abysmal lack of citizen participation. When one looks past the "smoke" of what government *should* do, the Act only required two public hearings.[35] As one who was Planning Director for the District of Columbia's Model Cities Program in 1969–1970, I can personally attest to the detailed body of knowledge regarding adequate public participation available to framers of the LGCPA of 1975.

Before one goes too far in pointing out the shortcomings of the Land and Water Management Acts of 1972 and the Local Government Comprehensive Planning Act of 1975, we must remember that no matter what, they were still a marked departure from the lack of regulation that preceded that period. It is almost impossible to turn around a process that took 120 years to develop in only four or five years. Had it not been for the extraordinary reform that took place during 1966–1970, the *rise* of the environmentalists *nationally* at the same time that an exceptional governor — Reubin Askew — was elected, this legislative package might never have been enacted. If the year had been 1982 instead of 1972, the climate would not have permitted such a strong turnaround, because by that time, after two major recessions — 1973–75 and 1979–82 — the public mood was more concerned about bread and butter issues such as jobs than about environmental protection. In fact, the development forces used the recession of 1973–75 not only to nix funding for the LGCPA, but to effectively urge legislators to consider loosening regulation in order to shore up the Florida economy during the years 1975–1980.[36]

What has that legislative package accomplished to date? As stated in the introduction, this book focuses on the issue of planning for growth in those areas designated for various intensities of the built environment. Therefore, one must look at developments of regional impact and the results of products stemming from the LGCPA. With respect to DRIs, Dr. DeGrove tells us that from July 1, 1973, through June 30, 1982, a total of 380 DRI applications were filed statewide. Of these, 243 were acted upon by December 30, 1981. Nine percent were approved without conditions, 7% were denied, and at most — 84% — were approved with conditions.[37] *What conditions?* Were they major or minor — changes of land use from commercial to residential, or word changes in the application? Analysis from the Department of Community Affairs data between 1982 and 1992 shows that of 568 DRIs acted upon, *1%* were approved, 3% were denied, and 96% were approved with conditions. Again, *what conditions?* Even for those approved unconditionally, the question remains, "Does their approval meet with good planning practice as per American Institute of Certified Planners standards?" The same questions apply to the LGCPA results. Dr. DeGrove's analysis shows that of 461 cities and counties required to prepare plans, as of 1978, 51 municipalities and 11 counties had submitted plans or even elements of plans for review and comment to the Division of State Planning. By mid-1982, 419 of 461 cities and counties had adopted plans that had been reviewed by the state, and finally, by 1984, all 461 units of government had adopted plans.[38] But Dr. DeGrove stated in 1984, "it is widely recognized that these plans vary greatly in quality. Furthermore, they have not been reviewed for consistency with regional and state plans that are themselves sufficiently clear and specific to be meaningful because such plans are not yet complete."[39]

By the early 1990s most of the local plans mandated by the 1985 legislation were complete, as were the regional and state plans. But were they any better than the plans produced by the 1975 legislation? The following chapters attempt to answer this question.

Notes

1. Rachel Carson, *Silent Spring,* Houghton Mifflin (New York), 1962.
2. Ian McHarg, *Design with Nature,* Natural History Books (New York), 1969.
3. For a complete and thorough review of the Cross Florida Barge Canal, the Miami Jetport, and The Everglades cases, see Luther Carter, op. cit. Note 2, pp. 82–116, 187–227, and 265–312.
4. For example, in this century, there have only been five commissioners of agriculture. B.E. McLin who served from 1901 to 1912, W.A. McRae who served from 1912 to 1923, Nathan Mayo who served from 1923 to 1960, Doyle Connors, who served from 1961 to 1990, and Robert Crawford, elected for the first time in 1991.
5. Some New Faces in Florida Could Defeat Incumbents, *New York Times,* October 17, 1970, p. B1.
6. *Time* magazine, May 31, 1971, pp. 18–19.
7. John DeGrove, *Land, Growth and Politics,* Planners Press, Chicago, 1984, p. 107.
8. Ibid., pp. 107–108.
9. Ibid., p. 108.
10. Fla. Statute, 380.05(2).
11. Fla. Statute, 380.06(11)(a–c).
12. Carter op. cit. at 2, p. 133.
13. Robert G. Healy, *Land Use and the States,* Johns Hopkins Press (Baltimore), 1976, pp. 111–112.
14. Even a city as large as Tampa (1970 population: 277,753) and Hillsborough County (1970 population: 490,275) were weak in terms of planning capacity. A special independent agency: the Hillsborough County Planning Commission, created by a local bill passed in the 1959 session of the legislature, was charged with planning, zoning, and subdivision review for not only the unincorporated portion of the county but all other units of local government, e.g., Tampa and the small communities of Plant City and Temple Terrace. In 1975, the commission had a staff of 95, with 30 holding professional level planning positions. However, of these 30 "professionals" only two held master's degrees in planning, which is the standard educa-

tional credential for planning professionals. The city of Tampa at that time did not have a planning staff as such, but there was a planning group in their urban redevelopment agency.

15. Department of Community Affairs Division of Technical Assistance: *The Local Government Comprehensive Planning Act,* October 1976, pp. 1–15.
16. Ernest R. Bartley, Local Government "Comprehensive Planning Act of 1975" *Florida Environmental and Urban Issues* (Ft. Lauderdale, September–October 1975), pp. 1–2, 13–15.
17. Department of Veterans and Community Affairs, 1980 Annual Report, February 1981.
18. Resource Management Task Force: Final Report of Governor Bob Graham, January 1980 (Executive, January, pp. 2, 3).
19. Daniel W. O'Connell "Growth Management in Florida: Will State and Local Governments Get Their Aid Together?" *Florida Environmental and Urban Issues,* April 1984, pp. 1–5.
20. Daniel W. O'Connell, "Legislating Quality Planning: The 1985 Local Government Comprehensive Planning and Land Development Regulation Act," *Florida Environmental and Urban Issues,* October 1985, p. 4.
21. Richard G. Rubino, "Growth Management Initiatives in Florida from Power Ranch to Saddlebrook," *Florida Environmental and Urban Issues,* January 1984, pp. 3–5.
22. Florida Statutes, Chapter 186.
23. Jack Osterholt, "Florida and State Plan: A Reflection of Growth and Change," *Florida Environmental and Urban Issues,* April 1985, pp. 8–11.
24. John M. DeGrove and Julian Conrad Juergensmeyer (Eds.), *Perspectives on Florida's Growth Management Act of 1985,* Lincoln Institute of Land Policy (Cambridge, Massachusetts), 1986, p. 7 (Introduction).
25. O'Connell, op. cit. Note 19, pp. 3–5.
26. See DeGrove and Juergensmeyer op. cit. Note 24, especially Chapters II, V, VI and VII.
27. Westi Jo DeHaven-Smith and Robert Paterson, "The 1986 Glitch Bill: Missing Links in Growth Management," *Florida Environmental and Urban Issues,* October 1986, pp. 4–6.
28. Section 163.3177(10)(a) Florida Statutes.
29. Brenda L. Valla, "Legislature Adopts Glitch Bill Amending Growth Management Act of 1985," *Growth Management Studies,* Vol. 1, No. 4, July 1986, p. 3.
30. Section 163.3177(10)(e) Florida Statutes.
31. John M. DeGrove and Nancy E. Stroud, "New Development and Failure Trends in Local Government Comprehensive Planning," *Stetson Law Review,* Vol. XVII, No. 3, p. 578.

32. Florida Department of Community Affairs, Informal Count of Local Comprehensive Plans, received as of December 31, 1990. Compiled by Kristia Kelly, January 14, 1991 .
33. See Carter op. cit. Note 2, pp. 132–133, Dealy op. cit. pp. 112–113 and DeGrove op. cit. Note 9, p. 118. Dr. DeGrove states that areas of critical concern designation are limited to 5% of the total area of the state *at only one time* as contrasted with Healy and Carter's analysis.
34. Department of Community Affairs, op. cit. Note 15, pp. 2–3.
35. Ibid., p. 12.
36. DeGrove op. cit. Note 9, pp. 125–128.
37. Ibid., p. 159.
38. Ibid., pp. 162, 172.
39. Ibid., p. 172.

Part II

The Rationale for Case Studies

During the period of 1984–1986 Florida's growth management tools were sharpened. The State Comprehensive Plan passed in 1985 provided the missing link of policy sets lacking in the Land and Water Management Acts passed in 1972. The Omnibus Growth Management Act assured a "top down" approach to managed growth, with consistency assured from the State Comprehensive Plan down to Regional Agency Plans, then down further to local plans prepared as a result of the 1985 Local Government Comprehensive Planning and Land Development Regulation Act (LGCPLDRA). Rule 9j5 adopted by the legislature in a slightly different form than submitted by the Department of Community Affairs provided some guidance for plan preparation in terms of standards and criteria.

In the previous chapter, I commented on the LGCPLDRA, especially R9j5, and the "glitch bill" by which the legislature adopted that rule. The package of new legislation passed between 1984 and 1986 will result in better local plans, but the question remains *"how much better?"* When one brushes aside all of the legalese, some glaring facts appear. First of all, no matter how tight state regulations are, those communities that take planning seriously will prepare optional elements as necessary, develop data to support recommendations, hold widespread, well advertised public hearings, and hold developer's feet to the fire to maximize concurrency *regardless of availability of funding*. Communities that don't take planning seriously will still turn out inferior products despite R9j5 no matter how much money is provided for plan preparation.

Second, by 1985 there were 524 professional planners in the state of Florida fully accredited by the American Institute of Certified Planners (AICP), compared to only 318 in 1978 when the AICP list first came out. If that wasn't enough, by 1985 there were over 5,000 certified planners nationally compared with only 2,300 in 1978. Planning has expanded vastly since the early 1970s, when Florida first began to deal with growth management issues. By 1985 computer applications, including Lotus 1-2-3 packages and the rudiments of Geographic Information Systems (GIS), were available to help compile and analyze data much faster and more accurately than in 1975, when computer applications for planning analysis were in their infancy. By the early 1990s GIS became available and data compilation and analysis could now be done much more quickly and thoroughly than ever before. What that meant is *professional planners did not need R9j5 to tell them how to prepare good plans.* The ability and analytical techniques were there. What was or was not there is the will of local government and its elite supporters (those who bankroll candidates for office) to accept planning as a community good and move forward to sincerely make certain that in the words of the "glitch bill," "local plans are not in conflict with the State Comprehensive Plan or appropriate Regional Policy Plan."

Another observation that had nothing to do with R9j5 is that no provision was made to mandate that local governments place a moratorium on zone changes while the community was preparing their plan update to conform with the 1985 legislation. This meant that communities found themselves making zoning decisions in 1991 with plans prepared as far back as the late 1970s. Was that commitment to planning? I don't think so!

Earlier I discussed briefly the problem of evaluating the quality of plans prepared under the 1975 LGCPA. Broad quantitative analysis is only the tip of the iceberg. What lies below the surface can only be detected by case analysis. The same phenomenon applies to products of the 1985 LGCPLDRA. For each of the plans submitted, what objections and recommendations and comments were made by the Department of Community Affairs? Which objections by DCA stuck despite the efforts of local government to get around them? How was conflict eventually resolved between DCA and units of local government?

This is why detailed case studies are needed to augment the broad-based quantitative assessment as to the effectiveness of both the 1975 Local Government Comprehensive Planning Act and its successor, the 1985 Local Government Comprehensive Planning and Land Development Regulation Act. Because as many as 15 years have passed since the adoption of the earliest products of the 1975 LGCPA, it will be much easier to evaluate the effect, as compared to the intent, of that legislation. However, this does not prevent examination of the process between plan preparation and DCA approval of plans coming as a result of the 1985 legislation. The first two cases are based in Tampa and Hillsborough County, and they examine the process of planning implementation as a result of the 1975 and 1985 legislation. The last two cases, situated in West Palm Beach and Gainesville, look at the 1975 and 1985 legislation, with emphasis on the latter. While in actuality all 461 case studies are really needed to fully depict the results of legislation adopted in 1975 and 1985, hopefully these four will present an initial, broad picture.

PART TWO

THE RATIONALE FOR CASE STUDIES

4 The Hillsborough County Solid Waste Site Selection Controversy: Struggling with a NIMBY

Introduction

Between the end of World War II and the late 1970s, Hillsborough County, which includes Tampa and two other smaller municipalities, underwent tremendous growth. In 1950, Hillsborough County had 220,000 residents. The population grew to 398,000 in 1960 and to 490,000 in 1970. By the late 1970s this county's population was approaching 650,000; in fact, the 1980 U.S. Census showed that 646,960 people resided in Hillsborough County. Between 1940 and 1960 most of this growth was contained in the city of Tampa, as that municipality, which had 108,391 residents in 1940, grew to 124,681 in 1950 and 274,970 in 1960. After 1960, Tampa's population leveled off, growing to only 277,753 in 1970, and actually decreasing to 271,523 by 1980. Growth in the remainder of Hillsborough County jumped from 185,000 in 1950 to almost 380,000 in 1980.

In 1957, a "reform" coalition composed of younger, junior chamber of commerce types, business elites, and middle management executive transferees helped to elect Julian Lane as mayor of Tampa. Part of Lane's reform agenda was to take zoning and subdivision regulation out of politics, and to accomplish this he pushed for a local bill in the state legislature that would remove zoning from under the City of Tampa's control and transfer it to an independent board. This bill, authorizing creation of the Hillsborough County Planning Commission, was introduced into the 1959 legislative session and was quickly adopted and signed into law.[1]

The legislation called for a ten member independent commission. Four members were to be appointed by the mayor of Tampa and confirmed by the city council. Four members were to be appointed by the Board of County Commissioners and one each from the small municipalities of

Temple Terrace and Plant City.[2] These cities were the only other incorporated municipalities in Hillsborough County besides Tampa, and both had populations under 10,000 in 1960. Funding was to come from the Board of County Commissioners, although in later years Tampa, Temple Terrace, and Plant City would make contributions for services rendered.[3] Originally this commission had only the power to prepare a land use plan including "majors plats, and charts for the orderly growth, and development of lands within the city of Tampa and lands in Hillsborough County."[4]

In 1961, amendments were enacted granting the Hillsborough County Planning Commission (HCPC) the right to review and make recommendations for all zoning adjustments, subdivisions, and lot splits in Tampa, Hillsborough County, Temple Terrace, and Plant City.[5] In 1963 another local bill was passed mandating that HCPC decisions on zoning and subdivision regulation could only be overridden by a *greater* than majority vote of the various units of county local government.[6] By 1964, HCPC had hired an executive director, ten professional planning staff members, and six support staff, and had an annual budget of $220,000. HCPC was actively promoted by the greater Tampa business community, as the Chamber of Commerce boasted that in addition to the city's fully developed infrastructure, good location, excellent weather, low wages, and a docile work force, the city had an independent planning commission "free from politics in order to plan for the orderly growth of greater Tampa."[7]

But HCPC was far from political freedom. As long as the staff processed zoning changes and subdivision plots to the satisfaction of the development community, they were left alone. On the other hand, when this body attempted to undertake real planning, they were turned back at every occasion. In 1962 HCPC staff presented its first countywide land use plan, only to have it rejected by the commission.[8] In 1967 HCPC staff tried again with a land use plan for 1990. It was adopted by its commission in a 6–4 vote, but rejected by both the Tampa City Council and the Hillsborough County Board of Commissioners. The same fate befell the 1974 plan for the greater Tampa Central Business District.

Concerned about runaway growth, the Hillsborough County legislative delegation, led by state Senator Guy Spicola, strongly pushed for the Statewide Comprehensive Planning Act in 1974. When that legislation failed, this same group passed a local bill in 1975 similar to the Statewide Local Government Comprehensive Planning Act of 1975, which did pass, except that (1) the required elements were tailored to fit Hillsborough County's needs, and (2) the plan had to be completed and adopted by all units of county government no later than December 31, 1977. The Statewide Act required completion and adoption of all plans by 1979.

Despite considerable opposition from developers, some HCPC Commissioners, and a few county agencies,[9] the plan known as "Horizon 2000" was completed in June 1977 and adopted by all four of Hillsborough County's local governmental units in December of that year. With over 800 pages of text, maps, and charts, Horizon 2000 contained the following elements: future land use, traffic circulation, sanitary sewer, solid waste, drainage and potable water, conservation, recreation and open space, housing, coastal zone protection, mass transit, port aviation and related facilities, utilities, and intergovernmental coordination. The final product (see Figure 4) was a broad policy-oriented Long Range Comprehensive Plan, which was criticized just before and immediately after its adoption as too broad and "fuzzy."[10] The key issue was, *given this plan, how well could it be applied to solving some of the county's "real world" problems, such as land use conflicts, inadequate vehicular circulation, overburdened utility systems, and worsening air quality?* The Hillsborough County solid waste controversy became the plan's first real test.

With Hillsborough County's rapid growth during the period of 1960–1980, coupled with the per capita increase in disposed waste experienced not only in the Tampa area but nationally as well, there was a desperate need for solid waste facility expansion toward the end of the 1970s. Despite advances in recycling and resource recovery technology, almost all sanitary engineers, elected officials, and citizen leaders realized there would still be a need for some type of landfill operation in Hillsborough County well past the year 2000. Knowing that nothing infuriates residents more than the expansion of an existing landfill or opening of a new one, elected officials were extremely cautious as to publicly announcing plans for new landfill capacity. But by the late 1970s, this issue could no longer be avoided. Existing landfills were reaching capacity and needed to be either closed or expanded. New landfill sites needed to be identified, selected, and then made operational. Tough decisions had to be made.

Central to the discussion in this case is (1) the role of Comprehensive Planning — especially that of the Solid Waste Element of Horizon 2000 — in the selection process, and (2) how the County Commission

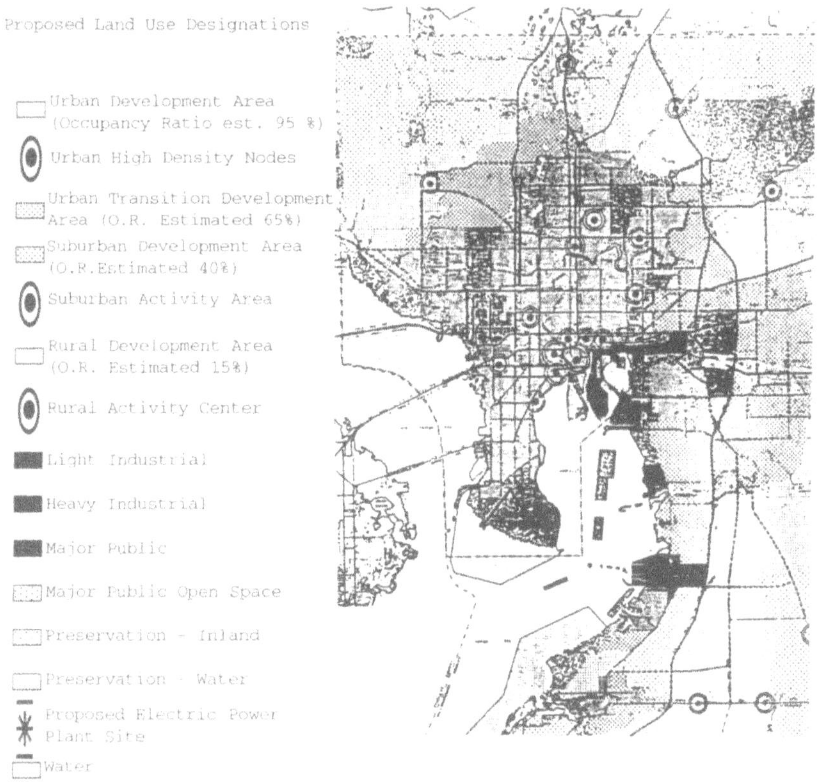

Figure 4. *Land use element, Hillsborough County. Generalized land use map year 2000. Land use and circulation: Horizon 2000 plan.*

weighed financial, engineering, and social science data with political considerations in determining site selection. The role of Horizon 2000 is critical because one of that plan's announced shortcomings by its opponents was that, as indicated beforehand, it couldn't be used because it was too broad, too general, and "fuzzy."

The landfill selection had to be made by the Hillsborough County Commission. Its members included Chairman Frances M. (Fran) Davin, Vice Chairman Robert E. Curry, and members Bob Bondi, Jerry Bowmer, and Joel Koford. All five commissioners were elected "at large."[11] Other players in this drama included the County Administrator William Tatum, the County Division of Public Utilities, which operates the solid waste landfills, engineering consultants for both the City of Tampa and Hillsborough County, citizen constituencies for both of the proposed landfill sites, and finally, the Hillsborough County Citizen Solid Waste Task Force, of which this author was a member.

Background

In June 1974, the Florida legislature enacted a law that, among other things, mandated that the Florida Department of Environmental Regulation develop extensive controls to ensure the environmental safety of future waste disposal operations anywhere in the state.[12] Furthermore, this legislation identified counties or groups of counties where resource recovery would be appropriate and mandated that the selected counties conduct extensive, detailed feasibility studies of resource recovery and submit resource recovery plans to the state. Hillsborough County was covered in this set of regulations.[13] As of June 1974, Hillsborough County was operating four landfills; two were in the areas of Ruskin and Gibsonton in rural south Hillsborough County. There was one landfill in the northwestern part of Hillsborough County in the rapidly suburbanizing Town N Country area, while the fourth landfill was located in northeastern Hillsborough at a location near the Mango Road interchange at Interstate 4 in Seffner. Seffner is a lower middle class semi-rural area which in 1978 was racially mixed, with 70% blacks and 30% white. The City of Tampa was operating an incinerator located in an industrial area near the port on McKay Bay and two landfills. The City of Plant City was operating its own small landfill in a rural area just inside that city's limits. In October 1974, the Florida Department of Environmental Regulation adopted the rules required by the state regulation referred to earlier here concerning waste disposal, facility design, and identification of areas for resource recovery study.[14]

During 1976–1977 several of the operating landfills in Tampa and Hillsborough County reached capacity and had to be closed. In May 1976 Hillsborough County closed the Northeast Landfill and immediately opened a 40-acre landfill on nearby Taylor Road. This landfill was expected to last only 4 years. It was quietly opened, and the residents, most of whom were low income workers or retirees living on social security, were not organized to effectively protest this action. Most thought that the County would only inconvenience them for a relatively short period after being given assurances in public hearings by the County Commission to that effect. Residents of the Taylor Road area were promised that the landfill would be abandoned by 1980 and the site turned into a regional park. Years later, when the landfill controversy was over, Pat Brantner, one of the Taylor Road residents, would say, "We really believed those bastards. How could we be so dumb?"

In July 1976, Hillsborough County closed the Gibsonton landfill. In August 1976 the city of Tampa received a second notice from the Florida Department of Environmental Regulation that it was in violation due to the pollution of Tampa Bay by the incinerator. Faced with massive cleanup costs, the City of Tampa opted to get out of the landfill business. It closed one of its two landfills in January 1977, and scheduled the other for closing in early 1978. The City of Tampa hired Dawkins and Associates, an engineering consultant out of Orlando, in March 1977 to develop plans for turning the incinerator into a resource recovery center and possibly finding a landfill site for joint City–County use, but outside of Tampa's city limits.

Hillsborough County government was keenly aware of a need for additional landfill capacity. In 1974, the engineering firm of Reynolds, Smith and Hill, of Jacksonville, recommended four transfer stations to serve unincorporated Hillsborough County. A transfer station is a facility where private and contract haulers can dump trash and garbage, which is then compacted on site and placed into containers for shipment by tractor-trailer to a landfill site. The use of these transfer stations minimizes not only the volume of solid waste but trips directly to the landfill site as well. After further study, Reynolds, Smith and Hill proposed that three transfer stations be constructed: one in Northeast Hillsborough, one in Northwest and one in the Southwest area of the county midway between Ruskin and Gibsonton. All three transfer stations were to be located in rural, undeveloped areas at least 2 miles from residential neighborhoods. In January 1977, Hillsborough County Government began to implement these recommendations by initiating construction of the Ruskin-Gibsonton transfer station and beginning the search for another landfill site — one that could last for 10–20 years.

In June 1977, Hillsborough County Government began to evaluate another of Reynolds, Smith and Hill's proposals — the use of Sydney Mine for a long-term landfill site. Sydney was located about 6 miles from the center of Brandon, an unincorporated urban place that had grown from a crossroads in 1950 to a community of suburban, middle class, single family homes housing almost 40,000 residents by the late 1970s. The site was potentially attractive — 3,500 acres in size, compared with less than 200 acres at Taylor Road. There were only 52 residents living within 1/2 mile of the site's borders. An old abandoned phosphate mine, Sydney had soil and subsoil characteristics that made it favorable for development as a solid waste landfill. With the Gibsonton-Ruskin transfer station operational by late 1978, the County's Division of Public Utilities estimated that the Taylor Road landfill would reach capacity in 1980. Given this situation, the Northwest landfill's permit was extended for one year for "high-rise" filling.[15] Also, in June 1977 plans were completed for an inter-local agreement between the City of Tampa, Hillsborough County, Temple Terrace, and Plant City. This agreement led to the formation of the Inter-local Management Committee, which was designated to conduct required resource recovery studies and plan for new landfill capacity as needed. Because by the late 1970s solid waste disposal was viewed by urban planners as a regional responsibility, the county Division of Public Utilities was charged with coordinating the Interlocal Management Committee efforts.

In August 1977, the City of Tampa entered into a consent decree with the U.S. Environmental Protection Agency to upgrade atmospheric emissions from its incinerator or cease operations by December 31, 1979. The City of Tampa then decided on advice from Dale Twatchman, its public works director, to phase out the incinerator and investigate its possible use as a resource recovery unit.

By late 1977 it was clear to all involved that Hillsborough County government had three options regarding solid waste disposal: (1) expand the Taylor Road facility to a possible 10-year life, then move to resource recovery; (2) acquire Sydney Mine as a new landfill site with a 20-year life; or (3) find another landfill site, a choice that up to that time had eluded them. If Taylor Road was expanded, the county government would incur the wrath of local residents, who were originally promised that the landfill would be phased out by 1980 and turned into a park. If Sydney Mine was selected, opposition from realtors and developers in Brandon, four miles west of the mine, would result. The developers viewed the area around Sydney Mine as a perfect location for single family residential development, and were not about to have their options foreclosed by a new landfill.

County Commissioners and civil leaders, especially those in the Brandon area, most of whom and their families were owners of single family homes, hoped that the forthcoming Horizon 2000 plan's Solid Waste Element would help resolve this dilemma. They were to be sadly mistaken.

The Horizon 2000 Solid Waste Element — Calling It As They See It?

Horizon 2000 was formally approved by the Board of County Commissioners on December 1, 1977. The Solid Waste element had been prepared by staff planners working with staff in the county's Office of Public Utilities. They had researched all relevant studies, including the 1974 Reynolds, Smith and Hill document. They had researched resource recovery, only to find that it would be years before an operation of that type could be conceived, funded, and built in Hillsborough County.

The Solid Waste Element's executive summary read as follows:

> The Solid Waste Element is primarily focused on aspects of waste management. As Hillsborough County has been mandated by the State of Florida to engage in a resource recovery plan and program, this element addresses the interim period between the present and such time as a resource recovery plan can be implemented. Realizing that the County as a whole is moving in the direction of resource recovery and that land adequate for sanitary landfills is becoming increasingly scarce, mobilization of waste has been specifically addressed. Attention has been focused on the utilization of transfer stations in the unincorporated County and the acquisition of a central disposal site. This site would provide long-term waste disposal capabilities for the County as well as provide a contingent disposal site for the municipalities. Even with a movement to resource recovery, land disposal of construction waste, and process residues will be necessary.[16]

After a brief summary of research and analysis findings, the Solid Waste Element went on to say that it was essentially an interim plan prepared to cover the period until resource recovery would be generally available, which would be anywhere between 1990 and the year 2000. The element called for the phaseout of Taylor Road, the last remaining landfill, by no later than 1982, but possibly as early as 1980. In the *Guidelines*

Regarding Solid Waste Disposal in Hillsborough County, the Solid Waste Element recommended identifying potential site-specific landfills that could handle solid waste for a minimum of 5 years after Taylor Road was closed, but preferably with a 20-year life, with the assumption that resource recovery would not be in full swing until just after the year 2000.

As far as a specific landfill site, the Solid Waste Element was indeed a bit "fuzzy." The element stated:

> In the 1974 Reynolds, Smith and Hill *2010 Study,* the Sydney Mine site was recommended as a central disposal site for all waste generated within the unincorporated areas of the County. This site consists of approximately 3,500 acres of undeveloped land; it is reasonably remote, yet is convenient to a major arterial highway (Highway 60). The development of this site into a sanitary landfill with waste flow from transfer facilities would give the County long-range disposal capability as well as providing a contingency disposal facility for the other municipalities.[17]

This was a "soft" recommendation that could be taken in many different ways. Proponents of phasing out Taylor Road as originally intended could say that Horizon 2000 recommended Sydney Mine as the longer-term alternative. But because of its vague language, opponents of Sydney Mine could claim that Horizon 2000 simply mentioned Sydney as one of any number of possible alternatives. County officials assumed that Horizon 2000's *mention* of Sydney as a possibility for a 20-year solid waste facility indicated that it at least deserved further study. County Commission Chairman Fran Davin told me in a 1978 conversation that the Commission would rather switch from Taylor Road to Sydney using Horizon 2000 and backup studies such as the Reynolds, Smith and Hill report to bolster their position "as long as residents around Sydney don't strongly object."[18]

Initially, it seemed that the county government was going with the Sydney Mine alternative. In March 1978, Hillsborough County hired the engineering consulting firm of Seaburn and Robertson to conduct a hydrological study of the Sydney Mine site. In July 1978 Seaburn and Robertson reported back to the County Division of Public Utilities that Sydney Mine was suitable from a hydrological point and recommended that it be used as a landfill. Decision time for County government was fast approaching as the Ruskin landfill was closed on July 31, 1978, with solid waste now diverted to Taylor Road, which now had only two years of useful life remaining on its existing site.

Brandon Residents Object to Sydney

While no formal announcement was made by the County Commission, word began to leak out that the Seaburn and Robertson study had endorsed Sydney as the new landfill site. Brandon civic leaders were enraged. Local realtors and developers had long coveted the area just west of Sydney Mine as prime residential property. Although Horizon 2000 had designated the area for 3 miles surrounding the site as "rural," Brandon realtors felt this provision could be changed by a plan amendment at the appropriate time.

The Brandon realtors, along with their associates in the Brandon Chamber of Commerce, started to mail out flyers to local residents as early as late August 1978. The flyers charged that the "Sydney Dump" would generate long lines of garbage trucks running down the community's main street (Route 60), spilling garbage and trash on their way to Sydney, and that their homes would be invaded by hordes of rats coming from the "dump."[19] These pamphlets were a bit misleading, to say the least. The proposed boundaries of Sydney were 4 miles from the built-up areas of Brandon's 40,000 residents and 5 miles from the initial landfill's center. The access route for trucks recommended in the Seaburn and Robertson and Reynolds, Smith and Hill studies was not Route 60, but Interstate 4 to Route 39 in Plant City and then south to Sydney — *completely bypassing Brandon.* Even the Horizon 2000 suburban residential area designation came only to within 3 miles of the Sydney site boundaries. As Taylor Road activists Tom and Pat Brantner would say later, "even the biggest rat couldn't run three miles from the landfill." However, the Brandon Chamber of Commerce didn't want anything that could possibly thwart development on the Brandon fringe, even if the danger was merely psychological.

The County Division of Public Utilities decided to hold an information meeting at Brandon High School on October 4, 1978. County staffers indicated their support for the Sydney Mine site to a crowd of over 800 persons in the high school auditorium. The crowd booed, then started throwing paper cups, then trash. A near-riot broke out and the Hillsborough County Sheriff's deputies had to be called in to restore order. The protesters were white and middle class and the deputies arrested no one, but instead pleaded for the crowd to "go home."[20]

Sensing a problem, the Hillsborough County Commission relied on the old adage "when in doubt, organize a task force to study the issue." A decision was quickly made to organize the Hillsborough County Citizen Solid Waste Task Force. This group was to consist of 16–20 citizens, who would come from various backgrounds: engineering, law, real estate, academia, business, and civic associations. Creation of this task force was announced by Dr. M. Frank Hersman, Director of the County Division of Public Utilities on October 18, 1978. The members of this task force, their mailing addresses, and professions, were in alphabetical order:

Name	Residence Address	Profession
Dr. Robert A. Catlin, AICP	Tampa	Associate Professor, Political Science University of South Florida
Mr. John Courson	Tampa	Comptroller Pepsi-Cola Bottling Co. of Tampa and President, Temple Terrace C. of C.
Mr. Julian Craft	Brandon	President, Craft Equipment Co.
Mr. Glenn Cross	Tampa	Vice President Shimberg Enterprises (Real Estate Development)
Dr. Tom Franques	Tampa	Professor, Department of Mechanical Engineering, University of South Florida
Ms. Martha Kjeer	Riverview, FL	President, Environmental Coalition
Dean Ed. Kopp	Tampa	Dean, College of Engineering, University of South Florida
Mr. Jerry Nichols	Brandon	Engineering Consultant and ex-President, Brandon Chamber of Commerce
Mrs. Pat Odiorne	Brandon	Member Alafia River Basin Board

Mr. Dave Puchaty	Tampa	District Manager, Southwest District, Florida Department of Environmental Regulation
Mr. William Peterson, Sr.	Tampa	President, Peterson Corporation
Ms. Marsha Rydberg	Tampa	Attorney (Gibbons, Zucker, McEwen, Smith Cofer and Taib Member, Junior League and Chairperson of its Solid Waste Task Force)
Mr. Bronson Thayer	Tampa	Vice President, Finance, Lykes Brothers, Inc. (Member, Greater Tampa Chamber of Commerce)
Mr. Larry Wagers	Tampa	Director of Operations, Hillsborough County School Board
Dr. Marshall Watkins	Plant City	Professor Emeritus, University of Florida, Co-Operative Extension Service
Mr. Dave Wolstrom	Sun City	Vice President of Construction and Land Development; W-G Corporation

Of the 16 task force members, *none were residents of the Taylor Road area, but three were from Brandon, including the past president of the Brandon Chamber of Commerce.* Two were from the development community, while three — Thayer, Courson, and Rydberg — were high-powered corporate types, with Rydberg representing Tampa's premier law firm. Four, including myself, were academics, and two represented public agencies, namely Florida DER and the Hillsborough County School Board. The rest could be described as citizen activists. This group was to have its initial meeting on October 24th.

The Task Force's Deliberations (See Figure 5)

The task force convened for the first time on October 24, 1978. Its purpose, according to a charter quickly drawn up by Hillsborough County administrators, was "to develop options for meeting the county's short range and long-term solid waste disposal needs."[21] Prior to the initial meeting, task force members received an information packet including a back-

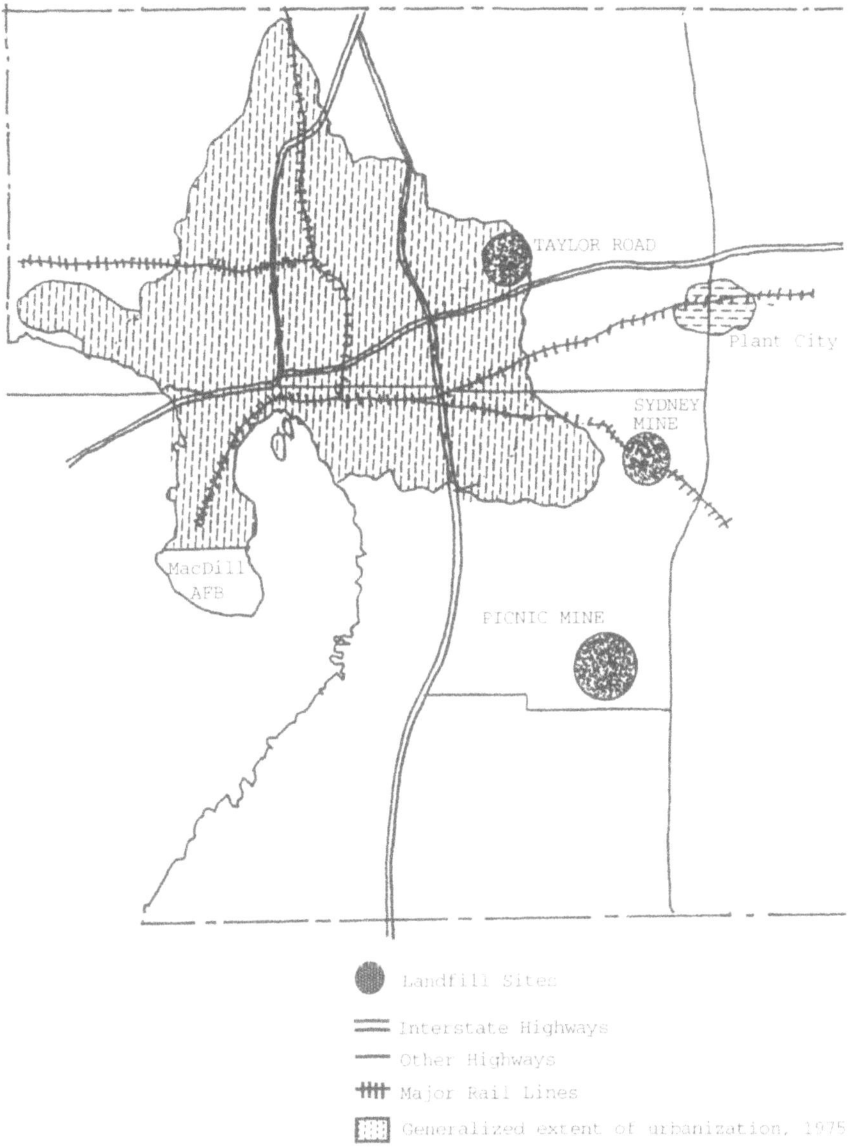

***Figure 5**. Landfill sites in Hillsborough County, 1975–1985.*

ground on the landfill controversy, a summary of the 1974 Reynolds, Smith and Hill study, and the Seaburn and Robertson hydrogeologic study completed in July 1978. The last page of that study concluded with: "*The site (Sydney Mine) is acceptable for solid waste disposal. It far exceeds the abilities of other sites in Hillsborough County and Florida to protect groundwater quality. It is basically a large impermeable basin that could retain any landfill leachate for collection and treatment."*[22] Other information items included statements on details of landfill operation, transportation to landfills, and resource recovery. Most of these pieces did not say anything really new. Horizon 2000 had already mentioned that resource recovery could not be fully implemented until just after the beginning of the 21st Century, that property located one mile from the border of a landfill was safe from odor and rodents so long as standard landfill management practices were adhered to, and that Sydney had the option of direct rail haul for dangerous waste products — something Taylor Road lacked completely. But the Seaburn and Robertson study was new. The fact that these consultants identified Sydney as the best site from the perspective of groundwater integrity was significant.

The meeting was convened by Commission Chairman Fran Davin, who thanked the members for agreeing to participate. After each member introduced themselves, Dr. Frank Hersman, Chief of the County Division of Public Utilities, introduced Len Joeris of his staff as the person who would be working directly with the committee, supplying us with all necessary data and information and answering questions. Joeris, an environmental specialist and a biologist by training, was a tall, lean, relaxed looking individual with a pleasant manner. A native of western North Carolina, Len was a strong environmentalist, who in the weeks and months to come would convince all of us of his fairness and desire only to see that the best site was the one selected — not the one that was politically acceptable.

The Task Force then proceeded with the election of chairman and vice chairman. After a brief discussion, Mr. Bronson Thayer of Lykes Brothers, Inc. was selected chairman and Mr. Jerry Nichols of Brandon was selected Vice Chairman. I could see then that the two sides of the Task Force would be (a) neutrals headed by Bronson Thayer, and including myself, and (b) the Brandon group, headed by Nichols, the ex-president of the Brandon Chamber of Commerce — a group totally opposed to Sydney, as could be readily ascertained by their opening statements. I asked, "Where are the Taylor Road representatives? There should be three of them, too, in order to even up the sides." My remarks were met by stony silence from the entire group, with some unfriendly glares on the part of the Brandon contingent. Bronson Thayer then adjourned the meeting.

At the second meeting, held on November 2nd, the Task Force organized subcommittees. Subcommittees were formed in the areas of finance, environmental impact, public information, landfill regulations, landfill operation, city-county cooperation, and social and community impact. I was chosen to head the social and community impact subcommittee. Presentations to the Task Force were made by Mr. Charles Hallenback, Director of Utility Operations for Hillsborough County; Mr. Larry Hall, Division of Public Utilities Hydrologist; Mrs. Joyce Morales of the Hillsborough County Environmental Protection Commission; and Mr. Keith Waller, senior planner with the Hillsborough County Planning Commission (HCPC). Hallenback discussed present county landfills in terms of their operational characteristics; Morales and Waller discussed the landfill siting controversy from the perspectives of their agencies. Hall's testimony was the most relevant; he stated that while vertical infiltration of contaminants into groundwater was not a problem at Taylor Road, horizontal movement of contaminants was. After the meeting, a few task force members and I had a private discussion with Mr. Hall, and were informed off the record that if Taylor Road were expanded there would be a clear danger of groundwater contamination of the wells for residents living within 1/2 mile of the expanded landfill's boundaries.

Between the second and third meeting, task force members took a field trip to look at Taylor Road, Sydney Mine, and a third site known as Picnic. I found Picnic to be particularly attractive because (a) it was an abandoned phosphate mine, like Sydney, with the same characteristics that made it feasible for landfill operations, namely, clay subsoils that made it virtually impossible for groundwater pollution; (b) it was located in the extreme southern portion of the county, about 20 miles from built-up urban or suburban areas, with only 16 dwellings within 1/2 mile of site boundaries, and 48 housing units and one shop within 1 mile of the site; and (c) it would be directly accessible from U.S. 41 and the new Interstate 75, which was scheduled for completion to Sarasota and Ft. Myers by 1982. There was even the possibility of rail haul. The only downside was that transportation cost could be almost double that for Taylor Road or Sydney Mine, according to Len Joeris.[23]

The third meeting was uneventful — more presentations by the Division of Public Utilities, with comments and questions by task force members. At the fourth meeting, held on November 16, progress was being made and a consensus began to develop. By that time the subcommittees had all met at least once. Public meetings were scheduled for January 11–22. I reported that my subcommittee would analyze social impact of the two sites using established Development of Regional Impact Criteria, even though the landfill selection was not a DRI, as it affected only one county (albeit a big one). Our intent was to present a report to the full task force by January 11.

By November 1978, the two major consultant groups — Seaburn and Robertson, and Dawkins and Associates — began to move in clearly different directions. In October, Seaburn and Robertson reaffirmed their recommendation on the use of Sydney Mine for the new landfill site in a presentation to the task force. But in November, Dawkins and Associates, who had been hired by the City of Tampa to find a landfill site outside of Tampa for joint city/county use, endorsed expansion of the Taylor Road site. Len Joeris told me after that meeting that the real reason for Dawkins' recommendation had little to do with environmental and social impact but had much to do with costs. Taylor Road would be the least costly of all alternatives to Tampa taxpayers because it was the closest site to Tampa, and with the landfill located well outside of the Tampa city boundaries, that city's elected officials would not have to face the wrath of irate citizens complaining about noise, odor, rodents, and increased truck traffic.

On November 30, the task force held its fifth meeting. By then, all subcommittees had held at least two meetings; my subcommittee had met three times earlier that month. Just days before this meeting, Brown and Caldwell, a sanitary engineering consulting firm with a nationwide operation, had signed a $100,000 contract with the county and the Interlocal Management Committee to conduct a detailed study on resource recovery. Also during November, the south county transfer station began operation and the county was notified by its Division of Public Utilities that Taylor Road had but 1 year of useful life remaining. On November 22, an agreement was reached between the City of Tampa and Hillsborough County to plan a joint solid waste operation for implementation as soon as the city incinerator shut down on December 31, 1979. The county would then operate all waste disposal facilities and Dawkins and Associates would conduct the necessary site-specific studies to determine expansion of the Taylor Road landfill. Seaburn and Robertson formally released their study to the public on November 29. The firm maintained their support for Sydney as the most feasible alternative.

The meeting itself consisted mainly of subcommittees reporting on their meetings. First to report was Social and Community Impact. Keith Waller of HCPC, and an ad hoc subcommittee member, stated that the solid waste and land use elements of Horizon 2000 were compatible with both sites under consideration. I contradicted Waller, pointing out that the solid waste element specifically mentioned Sydney Mine as a potential site as per the Reynolds, Smith and Hill study, while no mention of Taylor Road was made, although its selection as an interim landfill in 1976 was well known to HCPC staff. I also told the task force that the Taylor Road site was designated for "suburban" and "urban transition" land use, while Sydney was designed for "rural" use, as was the 3-mile area surrounding the site. "If the land use element means anything, Sydney is the site recommended by Horizon 2000," I stated. Waller retreated, stating that an HCPC staff report on the issue was forthcoming. The other subcommittee dealt with issues such as landfill operations, resource recovery, the need for public meetings, etc., which simply reinforced what was already known: with good landfill operations, the problems of rodents and odor could be limited to within 1/2 mile of the landfill's center, and resource recovery was at least 10 years away but more like 20. What was becoming clear was a need to make a decision soon. The Firestone Tire Company had recently recalled tens of thousands of its "Firestone 500" tires nationally due to defects. Now, hundreds of these tires were being dumped into Hillsborough County landfills. Tires present a special problem because unless they are split before dumping and compacting they come back to the surface, especially after heavy rains, bringing garbage and trash up with them. A tire splitter was ordered by the county, but as of the November 30 meeting it had not been received.

During January, public meetings were held in both the Taylor Road and Sydney areas. One by one the Taylor Road residents told in embittered tones that the county promised them a "temporary" inconvenience and then by 1980 a park. Now it was appearing to them that instead of a park, they would get a larger landfill. The tone of those meetings was one of resignation with suppressed anger. As one resident expressed, "you wouldn't be doing this to us if we weren't poor and black." The Brandon residents' position was generally one of righteous indignation — something to the effect of "how dare you do this to us."

By the end of January the facts were becoming clear to task force members. First of all, one of two decisions needed to be made: expand Taylor Road by 109 acres, making its utilization as a landfill possible through 1982 or initiate use of Sydney Mine on a 200-acre initial site, which could last through 1990. Over my protests, Picnic was ruled out by a majority of task force members due to high costs. The technical, social, and financial considerations clearly favored the use of Sydney. A description of each is in order.

Technical Considerations

First of all, from the many presentations made by staffers of the Division of Public Utilities, consultants, and other resource persons, it was clear that even with Taylor Road being expanded as far as possible, it still only had a life until 1982. With "high-rising," its useful life could extend only to mid-1984. High-rising Taylor Road would be extremely unattractive and further inconvenience area residents. At that point the county would have to look for another landfill site anyway, because all of the "experts" told the task force over and over again that resource recovery could not replace landfill needs until the year 2000, if then. Sydney, on the other hand, offered a 20-year maximum life. There were serious environmental problems with Taylor Road. During October and November, tests by the County Division of Public Utilities revealed numerous sinkholes and an unusually high groundwater table on the proposed expansion site. The only way to prevent contaminants from leaking into the groundwater would be the use of plastic liners, which would further increase costs, with no guarantee that the liners would be 100% effective. One wag on the utilities staff compared the liners with condoms, saying, "you know how those rubbers are, they always break at the wrong time." On the other hand, two-thirds of the Sydney site was free of sinkholes and subsurface problems. With fuel shortages becoming a real problem nationally by late 1979, the task force had inquired about the feasibility of rail haul not only for hazardous waste, but for shipment of compacted refuse to distant rural sites in nearby Polk County. County Utilities Division engineers told us this was a good idea, but of course, as stated earlier, only Sydney Mine had rail facilities. Also, we were informed that resource recovery worked best in industrial areas with rail facilities available. Again, Sydney had rail facilities, while Taylor Road did not.

Social Considerations

We found that 610 residents were within 1 mile of the proposed Taylor Road expansion and 548 were within 1 mile of Sydney Mine. However, within 1/2 mile of the Taylor Road expansion site there were 233 residences, compared with only 52 for Sydney Mine. The County Utilities Division staff told us that the 1/2 mile figure is more critical; within this area there could be serious impacts, such as rodent infestation and odor. Impacts were definitely more serious at Taylor Road than Sydney.

Our task force's Social and Community Impact Subcommittee, after hours of deliberation, came to the conclusion that no residential group liked landfills, regardless of income or occupational status. While this might seem to be a truism, several middle class white residents on the task force really believed that low income people didn't mind landfills — after all, the Taylor Road residents didn't complain when the county opened the landfill there in 1976. I had to remind the task force members that blacks didn't as a group complain about riding in the back of the bus until Rosa Parks decided one day in 1955 not to do so and the civil rights movement began. I reminded them that the urban riots of the late 1960s came about as a result of pent-up frustration and anger over police brutality, discrimination in employment and housing, and disparate treatment in the provision of public facilities. After weeks of arguing this point, the subcommittee came around and enclosed this statement, which appeared on page 89 of its report:

> *Community residents regardless of income group, housing type, or social class are equally opposed to landfills.* Lower income groups might not protest initially as much as middle or upper income residents, but this is due to lack of time to attend meetings, lack of organizational skills and discouragement at the prospect of a sympathetic governmental audience. However, lower income residents who are impacted by unwelcome prospects such as landfills internalize their frustration and the results are seen in alienation from the larger society, higher crime rates, mental/

> emotional illness, flight of upwardly mobile residents, and eventually urban disorders like those occurring in the late 1960s. Several leading sociologists subscribe to this viewpoint including David Cloward and Frances Fox Piven in *Regulating the Poor* (1968) and *The Politics of Turmoil* (1972), Robert Blauner in *Racial Oppression in America* (1971), and Alfred Kahn in *Studies in Social Policy and Planning* (1970).[24]

We felt that a dangerous precedent would be set if the county reneged on their promise to the Taylor Road residents. In 1976, the residents were promised that while the landfill was in use, the utilities department would provide landscaped berms, additional police protection, and traffic control. These promises were not met, and to further inconvenience residents by extending the landfill could be just the straw that would break the camel's back. From the January meetings with Taylor Road residents, I and other members of the task force felt that Taylor Road was a tinder box waiting to explode.

In terms of aesthetics, we noticed that Interstate 4 is the major eastern "gateway" to Hillsborough County and Tampa to motorists coming from the Northeast Seaboard via I-95. At that time, the Taylor Road Landfill was not terribly obvious. However, if the 109-acre expansion was approved — especially if it was high-rised — the landfill would be quite visible and the odor highly noticeable. Even with the best management practices possible, the site would be obnoxious and a turnoff to visitors. The task force, even those from Brandon, had to agree with this assessment.

We became convinced by late January that a decision to expand Taylor Road would be in conflict with Horizon 2000. Although Keith Waller, the staff representative to the task force from HCPC, would never admit it, I, Len Joeris, and eventually the task force majority clearly saw that the Taylor Road site was planned for suburban and urban transition development, which was a real possibility, as the site was only one mile from the I-4 and I-74 super-interchange. On the other hand, Sydney Mine and an area for a 3-mile distance from the site's boundaries were in the "rural" zone. (See Figure 4.)

Our major concern was that McDonald Elementary School, built in 1970 and with an enrollment of 900, was located only one-fifth of a mile from the proposed Taylor Road expansion. Additionally, truck traffic going to the landfill on a street facing the school would produce negative impacts, to say the least. While Buckhorn School was located 1.5 miles from the Sydney Mine site, truck traffic to the landfill would not pass anywhere near that school.

Financial Considerations

The finance subcommittee of the Task Force made its report to the full body on January 26, 1979. The group, working with Len Joeris, calculated that the cost per ton would be $4.88 at Sydney and $6.02 at Taylor Road. The difference was due to the fact that land costs were higher at Taylor, as were development costs, because the Taylor Road site had to be lined with plastic to prevent contaminants leaking into the groundwater supply. The cost at the Picnic Mine site was estimated at $11.15 per ton, with the higher cost due to excessive travel time from the county's populated areas to that site. Because of the high cost of the Picnic site, the finance subcommittee and the task force as a group decided not to give it any further consideration.

Len Joeris told the task force that actually land costs for the Taylor Road expansion were much higher than estimated. The expansion site was zoned commercial with shopping center potential because of the site's proximity to the I-4 and I-75 interchange and the existing Mango Road interchange on I-4. Both Glenn Cross of Shimberg Enterprises and Dave Wolstrom of W-G Corporation, the two developers on the task force, estimated privately that land cost for the Taylor Road site could double or triple even if condemnation suits were brought by county government in an attempt to overcome property owners' objections.

The Task Force Makes Its Recommendations

The task force took some straw polls in mid-January, and it seemed that at least six members, including myself, were leaning in favor of Sydney Mine as the recommended site. Four, including the three Brandon residents, favored Taylor Road, and the remainder were uncommitted. Word of our straw polls was leaked to the *Tampa Tribune,* which reported on January 16th that the task force was split 60–40 in favor of Sydney.[25] On February 9th the landfill task force of the Brandon Chamber of Commerce issued a public statement opposing Sydney Mine. The *Tampa Times,* on February 13, 1979, reported that a landfill at Sydney Mine would "hurt landowners."[26]

With virtually the entire county's attention turned toward the task force, Chairman Bronson Thayer felt that with all of the innuendo and speculation taking place it was now time to fish or cut bait. A meeting was scheduled for March 2, 1979 to take a final vote. After 4 hours of discussion by all 13 members present, the vote was 8 in favor of Sydney Mine and 5 in favor of Taylor Road. The 3 Brandon residents — Julian Craft, Jerry Nichols and Pat Odiorne — voted for Taylor Road, along with 2 others. I, along with 7 others, including Chairman Bronson Thayer, voted for Sydney Mine based on the weight of evidence presented by county government staff and private consultants.[27]

County staff in the Division of Public Utilities felt that more information was needed before they could endorse Sydney Mine as the preferred site. The tests were ordered for both sites on March 16, 1979, with the results due within 1 week. On March 19, 1979, Len Joeris, our staff contact and environmental specialist for the County Division of Public Utilities, wrote a detailed memorandum to his supervisor, Mr. Warren Smith. In this memorandum, Joeris stated that there were more sinkholes on the Taylor Road expansion site than first noted by the Dawkins and Associates report and that operational costs for Taylor Road could double because of the need to take extra precautions to prevent contamination of the groundwater supply. On March 20, 1979, Larry Hall, of the Division of Public Utilities, wrote an urgent memorandum to Pickens Talley, County Utilities Director, discussing liabilities of Taylor Road as pointed out by Joeris, and stating that "a decision on the expansion of the Taylor Road Landfill at this time may be premature." Talley then took both Joeris' and Hall's memorandums to County Administrator Bill Tatum on March 20, 1979.

That same evening I received a call from Len Joeris. Len, in a voice trembling with emotion, said "Bob, it's all over." "Tatum is bowing to political pressure. The county commission is going to vote tomorrow and it will be five zip for expansion of Taylor Road." I said, "Len, don't worry about it. We did our best. I hope our good commissioners will sleep well after this decision."

True enough, on March 21, 1979, the County Commission, in a vote announced by Chairman Fran Davin, chose 5–0 to expand the Taylor Road site, after publicly thanking the Citizen Solid Waste Task Force for their "hard work."[28] Commissioner Davin even went as far as to state on television that while the past promises to Taylor Road residents were not kept, the landfill will definitely close in 1982, be turned into a park and "then we will move to resource recovery." As I watched that evening news show, knowing that every consultant paraded before that task force said resource recovery would not be able to replace landfills until after the year 2000, if then, listening to Commission Chairman Davin, I kept thinking *"This is politics at its worst!"*

The Taylor Road Residents React

I and many other task force members were extremely upset at the County Commission for their decision to overlook our recommendation and choose Taylor Road for expansion without so much as a single dissenting vote or a rational explanation for the choice. One commissioner, in a haughty, patronizing voice said, when told that the expansion would pollute up to 100 wells of nearby residents, "We'll give 'em public water. That way they can take baths before they come to the courthouse to complain." In a *Tampa Tribune* article, I blasted the Commissioner's decision, calling it "short-term political expediency."[29] Pickens Talley, the County Public Works Superintendent who reported directly to County Administrator Bill Tatum, cautioned the Commission on rejecting the use of Sydney Mine totally because "resource recovery will not be implemented for years." However, Talley, being a good soldier, was not about to share with the news media the memorandums from Len Joeris and Larry Hall questioning the wisdom of expanding the Taylor Road site.

Now the Taylor Road residents were not to be patient and wait until being dumped on again without a protest. Led by Cam Oberling, a middle-aged housewife, who in her own words to me said, "I never protested anything before," and Tom and Pat Brantner, citizen leaders spread leaflets throughout the neighborhood. On May 13, 1979, the angry Taylor Road residents held a sit-in at the entrance to the landfill, blocking several dump trucks.[30] Roger Stewart, Head of the Hillsborough County Environmental Protection Agency, who was sympathetic to the Taylor Road residents but could do nothing because no environmental laws had been broken *yet,* advised the protest leaders to "get a lawyer and go to court."

On June 12, 1979, it was announced by the *Tampa Tribune* that the landfill expansion costs had increased by over $1 million from the estimates presented to the County Commission in March.[31] On June 27, 1979, the Florida Department of Environmental Regulation reported that water contamination of residents' wells from as far as 1/2 mile from the proposed Taylor Road expansion was highly likely.[32] On July 3, Mr. and Mrs. Brantner, Taylor Road area residents, filed suit to stop the landfill expansion. The Tampa Branch of the NAACP joined the Brantners in this suit. After the Florida Department of Environmental Regulation recommended on September 26 that Taylor Road be closed rather than expanded, the suit took on new life. Lawyers for the Brantners and the NAACP petitioned the U.S. District Court in Tampa for a restraining order barring acquisitions of additional land for the Taylor Road expansion. U.S. District Judge Ben Krentzmer issued a temporary injunction on October 8 in order to hear arguments for both sides. After the county pleaded that unless landfill expansion continues there will be a garbage crisis in Hillsborough County, especially with both the northwest landfill and the city of Tampa incinerator closing on December 31, 1979, Krentzmer lifted the injunction on December 17 and the landfill expansion, now raised to 120 acres, began. The Brantner's response to Judge Krentzmer's action was "the county acted like the kid who killed his parents and then asks the court for mercy because he's now an orphan. They started this mess by selecting Taylor Road for expansion. Too bad the judge agreed with them."[33]

The Taylor Road case continued to weave its way through the courts. On August 17, 1980, Florida Department of Environmental Regulations Secretary Jake Varn gave the county 18 months to close the Taylor Road expansion area, citing the same environmental problems pointed out over 18 months before by Len Joeris and Larry Hall.[34] In September 1980, two new County Commissioners, Joe Kotvas and Fred Anderson, were elected. Both men viewed themselves as populists, and they were enthusiastically supported by Taylor Road residents when they stated for the record that they had grave doubts about the landfill's expansion and would have voted against it if they had been in office at that time. Once elected, both commissioners said they would work to close Taylor Road as soon as possible.

On March 13, 1981, the county agreed to settle the Taylor Road suit by paying some 400 residents and their families over $3.1 million in damages and attorneys' fees.[35] With no progress made on resource recovery in over 2 years, the county turned that effort over to the city of Tampa on May 11, 1981. Taylor Road, after considerable high-rising and compacting, finally closed on October 31, 1984, and the county then opened Picnic Mine as the new landfill site. Known as the Southeast County Landfill, the Picnic site is bordered by a beautiful grove of orange trees that hide the landfill altogether from passing traffic on the road leading to the site. Tampa reopened the old incinerator as a resource recovery center on April 16, 1987, and by July 1991 it was disposing of 100% of the city's solid waste. Picnic has a useful life through 2005, and then, unless resource recovery is able to consume all of the county's solid waste, another landfill site will have to be chosen.

Analysis: Environmental Racism, Incrementalism, and the Impotence of Planning

Environmental Racism

The major focus of the Civil Rights movement in the 1960s was on discrimination in places of public accommodations, employment, housing, and enforcing the 15th Amendment for blacks in the South in the area of voting rights. This struggle has continued even into the 1990s, when a reluctant President George Bush signed into law the 1991 Civil Rights Act. In the 1990s, though, the focus of civil rights has shifted to another area: the battle against environmental racism (i.e., the tendency of public agencies to go against compelling technical and financial data and instead locate NIMBYs in areas populated by members of minority groups and poor people).

By the late 1980s, scholars began to compile a body of knowledge on this phenomenon.[36] Among them, this quotation from Dr. Robert Bullard, a sociologist from the University of California–Riverside and an expert in the sociology of environment and race, puts the entire issue in perspective: "Zoning, deed restrictions, and other 'protectionist' devices have failed to effectively segregate industrial uses from residential uses in many black and lower income communities. The various social classes with or without land use controls, are unequally able to protect their environmental interests. Rich neighborhoods are able to leverage their economic and political clout into fending off unwanted uses while residents of poor neighborhoods have to put up with all kinds of unwanted neighbors including noxious facilities. Public opposition has been more vocal in middle and upper income groups on the issue of noxious facility siting. The NIMBY syndrome has been the usual reaction in these communities. As affluent communities become more active in opposing a certain facility, the siting effort shifts to a more powerless community as opposition groups call for the facility to be sited 'somewhere else.' The Somewhere Else, USA often ends up in poor powerless minority communities."[37]

Bullard's observations are particularly relevant to this case. The County Commission, after reviewing the Reynolds, Smith and Hill study of 1974 and the recommendations of their personal consultants, Seaburn and Robertson, in mid-1978, were initially committed to move operations to Sydney Mine. Then when the middle class Brandon residents, riled up by half-truths from the realtor-backed Brandon Chamber of Commerce, objected in extremely strong and unprofessional terms, that same county commission voted unanimously to overrule its own hand-picked Citizen Task Force and move the dump to Taylor Road, located in the lower income Seffner area, which just happened to be predominantly black. Unfortunately, the ugly spectre of racism played a major role in this drama. It was interesting that no residents from Taylor Road were on the Citizen Solid Waste Task Force, although three Brandon residents were members, including the past president of the Brandon Chamber of Commerce. After the initial meeting I asked Dr. Frank Hersman, Director of the Division of Public Utilities and the organizer of the task force, why no Seffner/Taylor Road residents were members. Hersman first stated, "I couldn't find any qualified area residents." I replied "Frank, there are many well educated responsible and influential residents in the area." I mentioned the names of Dr. E.L. Bing, Mr. James Campbell, and his wife Doris Campbell, all black Taylor Road residents. Dr. Bing, a Hillsborough County School Administrator, had combined a career as an educator with cultivating his family's 80-acre orange grove. James Campbell, a college graduate, was an administrator with the Tampa Housing Authority, and his wife Doris held a master's degree and was a professor of nursing at Hillsborough Community College. After mentioning these names and qualifications to Hersman, he then replied, "Bob, we don't want to make this a black–white thing. We've worked hard to develop good race relations here in Hillsborough County, and if we put those three on the task force, we might wind up with fistfights in our meetings." I knew that, given those remarks, there was no need for further conversation on that issue.

If there had been three residents from Taylor Road, especially the ones I named, they would have balanced the emotionalism expressed by the three Brandon residents, and the task force could have moved in a rational and comprehensive manner to make the best choice — Picnic Mine. Leaving off the three Brandon residents would have served the same end. Then the task force would have been truly neutral. But in the final analysis, the notion that it was all right to appease the white middle class Brandon residents with three representatives on the task force while forbidding membership to any of the largely black and mostly lower income Taylor Road residents was racist to the core.

The County Commission and the task force as a group refused to consider the best site — Picnic Mine — because cost would be almost double that of Taylor Road. As it was, county government was forced to go to Picnic in 1984, almost 6 years after the controversy erupted. Actually, the costs of their decision were more expensive than going directly to Picnic Mine. The county spent over $5 million to acquire the expanded Taylor Road site and make it operational. Then the county, surprised by the angry Taylor Road residents' decision to go to court, had to pay out over $3 million to them in damages and attorneys' fees. Then in 1984, the county had to move to Picnic Mine anyway at an additional cost of $4 million for site acquisitions and preparations. *By making a short-term racist decision to save money, the county government actually caused their taxpayers to shell out up to eight million unnecessary dollars.*

The Faults of Incrementalism

Planning Theory according to Faludi[38] takes on at least three forms: rational-comprehensive, advocacy, and incrementalism. Incrementalism is a belief that because examination of all possible alternatives is impossible due to time, resources, and expertise, leaders must fashion incremental decisions or "one step at a time."[39] This form of planning theory tends to be preferred by public administrators because, having to almost always deal with day-to-day decision making, incrementalism clearly fits within the pattern of their time frames.

The Hillsborough County Solid Waste Controversy is a classic case of incrementalism. The County Commission only made decisive moves when they absolutely had to. A rational-comprehensive approach would have been for county government — especially the Division of Public Utilities — to send out strong signals to the Hillsborough County Planning Commission (HCPC) in early 1975 that in preparation of Horizon 2000 the solid waste element *must* indicate a path for the county and the city of Tampa to take to resolve the solid waste controversy before it became a major issue. The county already had by late 1974 (1) a mandate by the Florida legislature to come up with a plan for solid waste disposal, including resource recovery; (2) the knowledge that solid waste was becoming a problem due to rapid population growth and existing landfills reaching maximum capacity; and (3) the Reynolds, Smith and Hill study suggesting Sydney Mine as a long-term solution for landfill needs. Placing pressure on HCPC to utilize a truly rational-comprehensive approach would have resulted in a Solid Waste Element that identified Picnic Mine as the site of choice, given the technical, financial, and social costs involved with *both* Taylor Road and Sydney Mine. A Solid Waste Element naming Picnic Mine as the site of choice could have served as a county decision basis as early as January 1978. Then no meeting in Brandon would have been necessary, Seaburn and Robertson could have studied Picnic, showing its suitability as a landfill site, and the controversy would have been avoided. However, Keith Waller, the HCPC staff planner assigned to work with the Task Force's Social and Community Impact Subcommittee, told me during our deliberations that in the preparation of the Solid Waste Element, county staff in the Division of Public Utilities were generally uncooperative, unwilling to share information, and generally seemed to have the attitude that they were only talking with HCPC staff because they had been told by their superior officers to do so.[40] In the rational-comprehensive model of planning theory, the planner serves as an orchestra conductor, coordinating the work of various participants (including Public Utilities staff) to create a work of classic quality. When the public utilities staff essentially walked off the platform, Waller and the HCPC staff could not orchestrate. As a result, the Solid Waste Element did not give a reluctant and indecisive county government the direction needed to make a fitting and proper decision.[41]

The Impotence of HCPC

One of the major investigative themes of this book is *did long-range comprehensive planning as embodied by the Local Government Comprehensive Act of 1975 have major impacts on local government decision making with respect to mid-range development projects such as landfills?* In this case the author concludes that it did not. Even if the County Division of Public Utilities didn't fully cooperate with HCPC staff, whoever wrote the Solid Waste Element obviously knew that either Taylor Road's expansion or a new landfill at Sydney Mine would provoke controversy. Picnic Mine was abandoned by phosphate companies in the late 1960s, and its availability was widely known to planners, real estate developers, and public administrators. It was unfortunate that Picnic Mine was overlooked by HCPC staff in 1976 when Horizon 2000 was under preparation, because in 1978 my associates on the task force and I noticed it after we asked Len Joeris, "Are you sure there aren't potential landfills sites far out into the rural areas of southeast Hillsborough?" Although the task force, taking direction from its chairman, limited the choices to Taylor Road and Sydney Mine, in February 1978 several task force members, including myself, were quoted in the *Tampa Tribune* as favoring Picnic Mine.[42]

Having blown it on the Solid Waste Element, HCPC *still* could have used the rational-comprehensive approach in early 1978 to carefully examine all possible sites and then recommend Picnic Mine — which after all was where the county went anyway after Taylor was used up. Keith Waller never did get back to my Task Force Subcommittee with a HCPC recommendation after he promised one at the November 29, 1978 meeting. A strong HCPC recommendation on Picnic or even Sydney or would have been a sign of the independent Planning Commission's viability and evidence that HCPC *could* do effective planning and zoning recommendations — or walk and chew gum at the same time. However, HCPC chose to play it safe and stay out of the limelight and away from controversy. By doing so, not only did HCPC create its own credibility gap, but more importantly, the struggle to use rational planning in local government decision making was set back considerably.

Conclusion

Hillsborough county government and the independent Hillsborough County Planning Commission could have utilized the rational-comprehensive process to identify a new landfill site as far back as 1974, given population growth, existing landfill capacity, and consultant studies. However, both groups chose incrementalism, or the line of least resistance. Initially the game plan was to use Taylor Road, opened in 1976 without protests from nearby residents until 1980, then shift to Sydney Mine. However, when the Brandon residents strongly protested the potential choice of Sydney, the County Division of Public Utilities organized a Citizen Task Force to study the issues and then make recommendations. The task force recommended 8–5 that Sydney be used as the new landfill, and I personally recommended the choice of Picnic. The county commission ignored the task force recommendation and unanimously voted for an expanded Taylor Road, assuming all the while that (1) the Taylor Road residents wouldn't protest, and (2) the cost of going to Picnic would be too high due to extended travel.

Taylor Road residents not only protested, but went to court. They formed some unexpected allies in the Florida Department of Environmental Regulation and two new county commissioners — Joe Kotvas and Fred Anderson — who sided with them. The result was a 1981 consent decree in the U.S. District Court, which cost the county $3.1 million in payments to the Taylor Road residents and attorney fees. Finally, in 1984, Taylor Road had to close. The county commissioners' dream of using Taylor Road and then going to resource recovery turned out to be a nightmare, because even though 22 metropolitan areas across the nation had full service resource recovery operations in place by 1984, Hillsborough County was not one of them. The county had no choice but to go to Picnic Mine, after spending over $8 million of taxpayers' money to expand Taylor Road and buy off the area residents and their attorneys.

The Hillsborough County Planning Commission's failure to use the rational-comprehensive approach to first develop a Solid Waste Element for the Horizon 2000 Plan in 1977, and then in 1979, and failing to provide leadership for landfill siting by the County Commissioners, was proof of that body's impotence and the accompanying failure of long-range comprehensive planning to exert meaningful influence in critical mid-range decision making and subsequent implementation.

In summation, the real losers in this drama were not necessarily the Taylor Road residents. Some 400 of these residents had their wells replaced by a public water system at no cost to them. These residents also received free water so long as they or their heirs occupied their homes. The losers were Hillsborough County taxpayers, saddled by an $8 million bill due to racism and incompetence. The biggest loser of all was comprehensive planning and the Hillsborough County Planning Commission, which embodied its spirit and purpose.

Notes

1. Laws of Florida 59-1363 (1959).
2. The original legislation called for only three members to be appointed by the Hillsborough Board of County Commissioners. This number was raised to four in 1961 (Laws of Florida 61-2262).
3. Op. cit. Note 1, p. 1649 (Laws of Florida).
4. Ibid., p. 1652.
5. Laws of Florida 61-2262 and 61-2264.
6. Laws of Florida 63-1407.
7. Annual Report, Tampa Chamber of Commerce 1964, p. 27.
8. "Citizens, Landowners Resource Land Plan" *Tampa Times,* November 12, 1962.
9. For example, see Gel Klein, "Senator Says Land Plan Sabotaged," *Tampa Tribune,* May 22, 1977, p. B1 and Jon Pech, "Poe Says Horizon Plan Neglects Human Sources," *Tampa Times,* August 4, 1977, p. B1.
10. "Member Pushes Planning Panel's Revamping," *Tampa Tribune,* December 30, 1977.
11. "At Large in Hillsborough County, Florida" means that while an individual must reside within a defined geographical area of the given governmental jurisdiction, that person is subject to the entire electorate.
12. Laws of Florida, Chapter 74-403 Part 4.
13. Ibid., Part 4 and 5.
14. Regulation 17-7. Florida Administrative Code (F.A.C.).
15. "High-rise" is a technique where garbage is dumped, covered, and compacted, then the process is repeated directly on top of the compacted surface. The result is a hill that can and does rise hundreds of feet above ground level.
16. Executive Summary, *Horizon 2000 Plan Solid Waste Element,* Hillsborough County Planning Commission, December 1, 1977.
17. Ibid., p. 14.
18. Conversation with Hillsborough County Commission Chairman Fran Davin, September 8, 1978.
19. Brandon Chamber of Commerce pamphlet, August 30, 1978.
20. Jeff Kahs "Brandon Residents Say No Landfill," the *Tampa Tribune,* October 5, 1978, p. B1.
21. *Charter for the Hillsborough County Citizen Solid Waste Task Force.* Hillsborough County Administrator's Office, October 20, 1978, p. 1.
22. Seaburn and Robertson, Inc. *Hydrogeologic Assessment of the Proposed Sydney Landfill Site — Stage I,* p. 78, July 1978.
23. *Tampa Tribune* reporter Greg Tozian talked to me several times after the Task Force field trip and my response on each occasion was that Picnic

Mine was the best alternative for a new landfill site. He mentioned my views on Picnic in an article published on February 22, 1979 in the East Hillsborough section of the *Tampa Tribune* entitled "Task Force Subcommittee Recommends Investigation of Picnic Mine as Landfill Site," p. 5.

24. Report of the Subcommittee on Social and Community Impact, *Citizen Solid Waste Task Force,* February 1, 1979, p. 89.
25. Greg Tozian. "Landfill Task Force Set for Recommendations," *Tampa Tribune,* January 16, 1979, p. B1.
26. Greg Tozian. "Brandon Landfill to Cost Taxpayers Millions," *Tampa Tribune,* February 13, 1979, p. B1.
27. Greg Tozian. "Sydney Site Advocated for Landfill," *Tampa Tribune,* March 3, 1979, p. A1.
28. Greg Tozian. "Dump to be Expanded: Rate Hike Seen," *Tampa Tribune,* March 22, 1979, pp. A1, A3.
29. Greg Tozian. "Neighbors Call Landfill Choice a Slap in the Face," *Tampa Tribune,* March 23, 1979, p. B1.
30. Greg Tozian. "Taylor Road Residents Block Dump Entrance," *Tampa Tribune,* May 14, 1979, p. B1.
31. Greg Tozian. "Dump Costs to Rise," *Tampa Tribune,* June 20, 1979, p. B1.
32. Greg Tozian. "Pollution of Neighbors Wells Likely Due to Dump Expansion," *Tampa Tribune,* June 27, 1979, p. B1.
33. Interview with Mr. and Mrs. Thomas Brantner, Taylor Road residents, January 16, 1980.
34. Greg Tozian. "DER Gives Taylor Road Dump 18 Months," *Tampa Tribune,* August 18, 1980, p. B1.
35. Interview with Mr. Al Munger, public affairs officer, Hillsborough County Division of Public Works. October 28, 1991. The damages included $2 million for public water provision for 400 residences and business, $800,000 in acquisition costs for 15 properties and cash awards of over $400,000 for attorney fees.
36. See Robert Bullard and Beverly Hendrix Wright, "Environmentalism and the Politics of Equality: Emergent Trends in the Black Community," *Mid American Review of Sociology,* 1987, Vol XII No. 2: 21–38. Robert Bullard, "Solid Wastes and the Black Houston Community," *Sociological Inquiry,* Vol. 53 No. 2/3, Spring 1983. Robert D. Bullard and Beverly Hendrix Wright, "Blacks and the Environment" *Humboldt Journal of Social Relations,* Vol. 14 (Summer 1986) 165–184, and Robert Bullard, *Dumping in Dixie: Race, Class and Environmental Quality,* Westview Press (Boulder, Colorado), 1990.
37. Robert D. Bullard. "Environmental Blackmail in Minority Communities," Unpublished Paper presented January 11, 1990, pp. 7–8.

38. Andreas Faludi (Ed.). *A Reader in Planning Theory*, Pergamon Press (New York), 1973.
39. Charles Lindbloom. "The Science of Muddling Through" in Andreas Faludi, *A Reader in Planning Theory*, Pergamon Press (New York), 1973, pp. 151–170.
40. Interview with Keith Waller, HCPC Senior Planner, February 11, 1978.
41. The same question "Why didn't the solid waste element note Picnic?" was posed to Len Joeris and Larry Hall of the county staff. Their response was that they didn't want to discuss anything about the Division's relationship to HCPC. Interviews with Len Joeris and Larry Hall, February 11, 16, 1979.
42. Tozian, op. cit. Note 12.

5 Old Hyde Park Village: An Example of Transactive Planning

Introduction

In the first case study we saw how county government bowed to pressures from the middle class Brandon community, egged on by the local chamber of commerce, many of whom were part of the development alliance, and expanded a landfill in the low income, majority black Seffner–Taylor Road neighborhood.

In this case, the quality of planning and the character of the community were quite different. The setting is the Hyde Park neighborhood of Tampa — predominantly white, middle class, professionals, and small businesspersons, well educated, articulate, and extremely well organized. The plan in question was not necessarily Horizon 2000, but a mid-range neighborhood development scheme. It was drawn up by neighborhood residents working with HCPC staff and adopted by the city of Tampa in June 1976, at the same time Horizon 2000 was being prepared and 3 years before an out of town development group proposed a mammoth shopping center on a site the plan had designated specifically for low intensity neighborhood commercial.

In the previous case, we found that initially the intervention chosen by Hillsborough County government, namely, expansion of the Taylor Road landfill, proceeded exactly as planned, although the county, not anticipating the militancy of local residents, wound up paying millions of dollars in what turned out to be a major judgement error. In this case, the intervention of the developer was considerably modified as the process unfolded. The role of planning was quite different in this case in terms of its initial impact on the process and the eventual outcomes.

Background

In 1884, Henry Plant extended his railroad west from the Atlantic Ocean to Tampa, which at the time was no more than a sleepy fishing town with fewer than 1,000 residents. At the end of his rail line, Plant built in 1891 the large ostentatious Tampa Bay Hotel as a mecca for wealthy northern tourists. Located just west of Tampa's downtown on the banks of the Hillsborough River, the Tampa Bay Hotel became the center of Tampa's social elite, especially after its name appeared in newspapers all over the world as headquarters for Major General William R. Shafter, commander of the Expeditionary Force forming to go to Cuba (which included Lt. Col. Theodore Roosevelt's Rough Riders) during the Spanish-American War.[1] In 1901, O.H. Platt, from Chicago, subdivided 20 acres just south of the Tampa Bay Hotel into 1/2- and 1/4- acre lots. He named this subdivision Hyde Park, after his Chicago neighborhood, which had been the site for the world famous Columbian Exposition in 1893 and 1894. Between 1901 and 1929—the beginning of the Depression—several mansions for the wealthy were built in Hyde Park and it became a fashionable address for prominent Tampans. During the 1920s land boom, many of the unbuilt 1/2- and 1/4-acre parcels were further subdivided into lots as small as 5,000 square feet, or 1/8 of an acre, and smaller bungalows were built alongside the earlier mansions. Also, with no zoning in place, apartment buildings and grocery stores were built on scattered sites, so that by 1930 Hyde Park featured a mixed land use pattern.

The Depression halted residential construction, and during World War II many of the older mansions and smaller bungalows were cut up into apartments as Tampa's population swelled due to employment growth in the city's shipyards and military bases. After the war, affluent and upwardly mobile middle class residents abandoned Hyde Park for the suburbs. These residents were replaced by lower income workers and their families. A small enclave of 4 blocks in the neighborhood's northwest corner that before World War II housed the servants of wealthy Hyde Parkers now began to slowly expand as a lower income black ghetto. With this shift in population, city services declined, white flight continued, and by the late 1960s Hyde Park had generally deteriorated.

But things changed quickly. In the early 1970s, with the Tampa area growing by leaps and bounds without corresponding improvement in roadway capacity, the suburban commute became difficult and expensive. Hyde Park, which was within walking distance of downtown and the University of Tampa, now housed in the old Tampa Bay Hotel, became attractive once more. Young urban professionals began purchasing the old charming mansions and renovating them. Older apartment buildings and bungalows alike became new office settings for physicians, attorneys, realtors, and insurance people. Old grocery stores now became wine and cheese shops, and old taverns that formerly catered to laborers now became fern bars.

By the mid-1970s, Hyde Park had a majority of "yuppies," but a sizeable minority of elderly and lower income residents from the old days remained as well. While renovation was extensive, even during the 1973–74 recession there remained significant pockets of substandard housing, vacant storefronts, and weed-infested lots. Speculators began to purchase run-down buildings and vacant lots, waiting for values to go up, but all the while refusing to repair the buildings or mow the lots. In 1974, the Crosstown Expressway was built through the middle of Hyde Park, removing many residents and their older multifamily housing units. Construction of this expressway also cut through the fabric of Hyde Park, causing disruption and anxiety. Finally, the newer residents formed the Hyde Park Civic Association and petitioned the Hillsborough County Planning Commission to help them draw up a neighborhood plan to guide future growth and development.

The Hyde Park Plan

In 1974 the Tampa City Council, spurred on by the Hyde Park Civic Association, asked HCPC to prepare a redevelopment plan for the central core. Rather than a single plan, separate plans were prepared for Tampa's downtown, Seddon Island (which would later become the planned community of Harbour Island), Davis Islands (the 1926 development by speculator "Doc Davis" mentioned in Chapter 2), and Hyde Park. The Hyde Park plan was prepared as a mid-range development scheme reminiscent of Martin Meyerson's "middle range bridge."[2] It showed land use on a site-specific block-by-block basis (see Figure 6). Its Transportation System Plan showed existing and proposed arterials and collectors, existing bus routes, and a mass transit corridor in the median of the Crosstown Freeway. The plan was prepared with maximum public input, including questionnaires mailed to 400 residents, over two dozen public meetings, and four public hearings. The printed document stated that the plan's purpose was "to provide an official guide to the future development of the

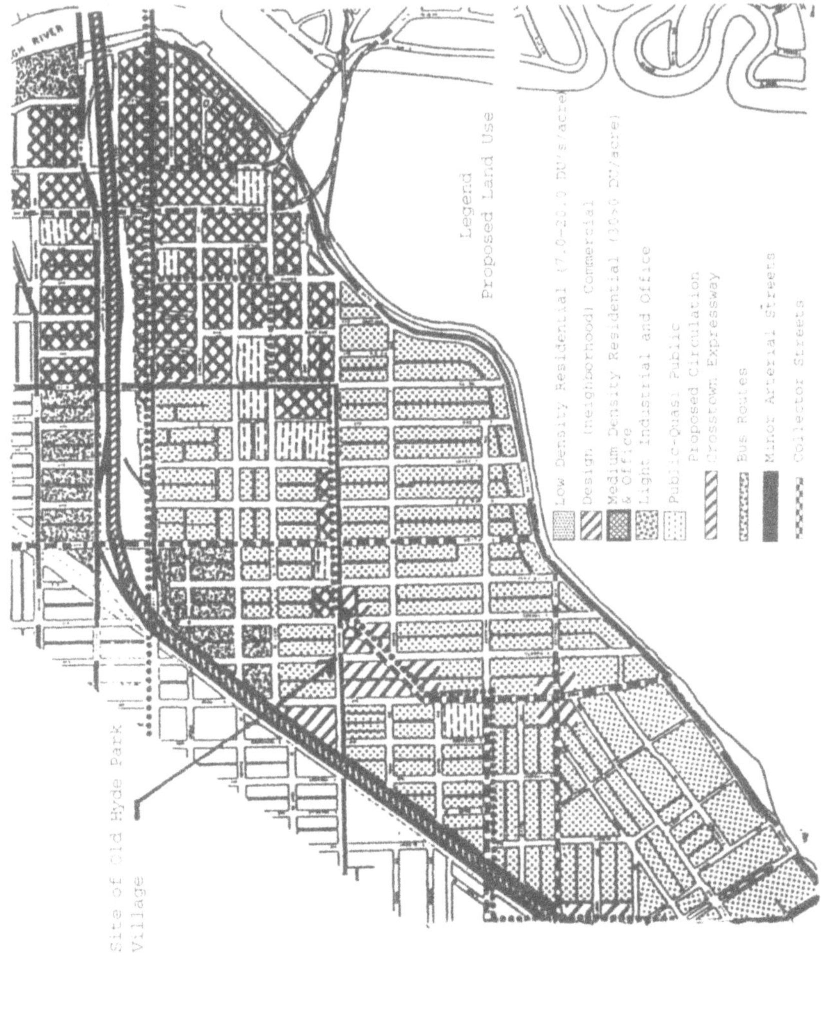

Figure 6. The Hyde Park Plan, 1975.

neighborhood for use by the Hillsborough County Planning Commission Staff and members, the mayor, the City Council, and other concerned governmental agencies, residents, property owners, and businesspeople of the neighborhood, and private organizations concerned with neighborhood improvement."[3] A statement that would have tremendous significance a few years later read *"The Plan will also provide an officially approved reference to be used in connection with action on various City development matters as required by law."*[4] What was being said here was that any changes of zone or public acquisitions must be in conformance with this plan.

This plan was adopted by the Hillsborough County Planning Commission and the Tampa City Council in June 1976.[5] When the Tampa City Council passed its version of Horizon 2000 after considerable revision on June 28, 1979, the Hyde Park Plan of 1976 was directly incorporated into it with only minor land use changes — those being on the periphery.

Therefore, by 1979, Hyde Park had a neighborhood plan, which was part of a long-range comprehensive plan. It was site-specific in terms of proposed land use. Developers desiring to build in Hyde Park had not only existing zoning to go by, but a plan that showed proposed uses on a parcel-by-parcel basis. This was in direct contrast to suburban Tampa and most other locales in Florida, where potential developers had to fly blind, as the community might not even have a zoning ordinance or a broad long-range comprehensive plan, to say nothing of something so specific as a neighborhood development plan. The bottom line was that anyone who wanted to develop properties in Hyde Park knew as early as 1976 exactly where they stood with respect to long-range policies, mid-range development land use categories, and existing zoning. Would developers take advantage of this clear picture and buy land that was not only zoned but planned for their intended uses? Not in this case!

The Origins of Old Hyde Park Village

In March 1971, with the gentrification of Hyde Park in full swing, local Tampa realtor James Burt purchased several structures in the vicinity of Swann and Snow. These buildings housed small restaurants, high-quality clothing boutiques, and professional offices. However, Burt was set back by the 1973–74 recession, and finally in March 1979 sold his properties to the Amlea Corporation, a firm based in Toronto that specialized in large shopping center development. Amlea's cost for the property was $8 million. The Swann–Snow intersection was zoned "neighborhood commercial" and the Hyde Park Plan called for "design commercial." This designation was described in the Hyde Park Plan as follows.

> *Commercial areas* in Hyde Park will be special in their orientation and design. Their purpose is to provide for specifically site-planned commercial areas which meet a defined service need such as neighborhood pharmacies, barbershops, or specialty uses such as pastry or flower shops.... All commercial development will have to submit a site plan showing design compatibility, landscaping building, and parking placement. Hyde Park is too special an area to allow strip commercial development to further congest its heavily trafficked streets, or to build structures that would destroy the neighborhood character.[6]

Nothing could be clearer. The Swann–Snow intersection was planned for low-intensity commercial uses similar to the existing ones and compatible with the surrounding neighborhood character as embodied by the one- and two-story single-family residences that dominated the area. Burt's plans for his properties were not well-known, except that some level of commercial activity was anticipated. After the Hyde Park Plan was adopted by all governmental agencies, the expectation of neighborhood residents was that Burt would develop the properties in accordance with the plan — a few additional single-story shops complementing existing ones, perhaps a small two-story building with shops on the ground level and offices above. The older shops would be rehabilitated and there would be a unifying architectural theme tying together the new buildings with older ones. Some lots would be reserved for parking. As befitting a "neighborhood" shopping center, the maximum square footage for the entire development would not exceed 100,000 as per recognized standards promulgated by the Urban Land Institute.[7] When the Amlea Corporation purchased James Burt's properties in Hyde Park, local residents felt that, especially with this firm coming from such a well-planned city as Toronto, they would, with Canadian reserve, respect the zoning in place — the Hyde Park Plan — and propose a low-profile, high-quality center similar to those evolving in places such as Miami's Coconut Grove and Palm Beach's Worth Avenue.

On June 28, 1979, the Tampa City Council passed its revised version of Hillsborough County's Horizon 2000 Plan. The Hyde Park Plan, slightly revised in 1977 when Horizon 2000 first came out, was now incorporated as part of Tampa's Comprehensive Plan. However, one week later, on July 2, 1979, Amlea announced plans for a 350,000-square-foot shopping center anchored by Jacobson's, an upscale junior department store. There would be an additional 40,000 square feet of offices and 100 condominiums on the site. Parking would be provided by lots and garages operated by the City of Tampa, with parking meters and gated entrances. Amlea would need not only a zone change but a plan amendment for this development, which was more on the scale of a suburban "community" shopping mall than the quaint "specialty" center called for by the Hyde Park Plan and existing zoning.

Neighborhood residents were furious. They realized that Amlea had pulled the basic developer's trick — *buy cheap, then get rezoning*. The fact that a meticulously drawn plan, prepared in cooperation between HCPC staff and neighborhood residents and adopted by the Tampa City Council three years beforehand, and reaffirmed when Horizon 2000 was adopted only one week before on June 28th — a plan still calling for low-intensity development on the site — meant absolutely nothing. Amlea's plans called for four three-level buildings each 40 feet high, with the first level devoted to shops, and the second and third levels to parking or residential units. There would be an additional three-level parking garage and a 50-foot building over 200 feet in length that would house four floors of retail office and residential condominium units (see Figure 7). With surrounding single-family residences being one and two stories high, and with no structure exceeding 25 feet in height, this center would be a radical physical imposition on the neighborhood. Hyde Park residents who just happened to be civil engineers working for consulting firms or city/county government quickly calculated that this center would generate about 20,000 trips per day based upon the nationally recognized Institute of Traffic Engineers (ITE) standards and criteria. The proposed development would be located at the intersection of a collector and a minor arterial, both of which were operating at service level "D" according to data compiled in 1975 for Technical Reports I and II of the Hyde Park Plan. The additional trips generated by the Amlea proposal would further undermine the level of service on these roadways, as "C" is considered by traffic engineers to be adequate and "D" inadequate.

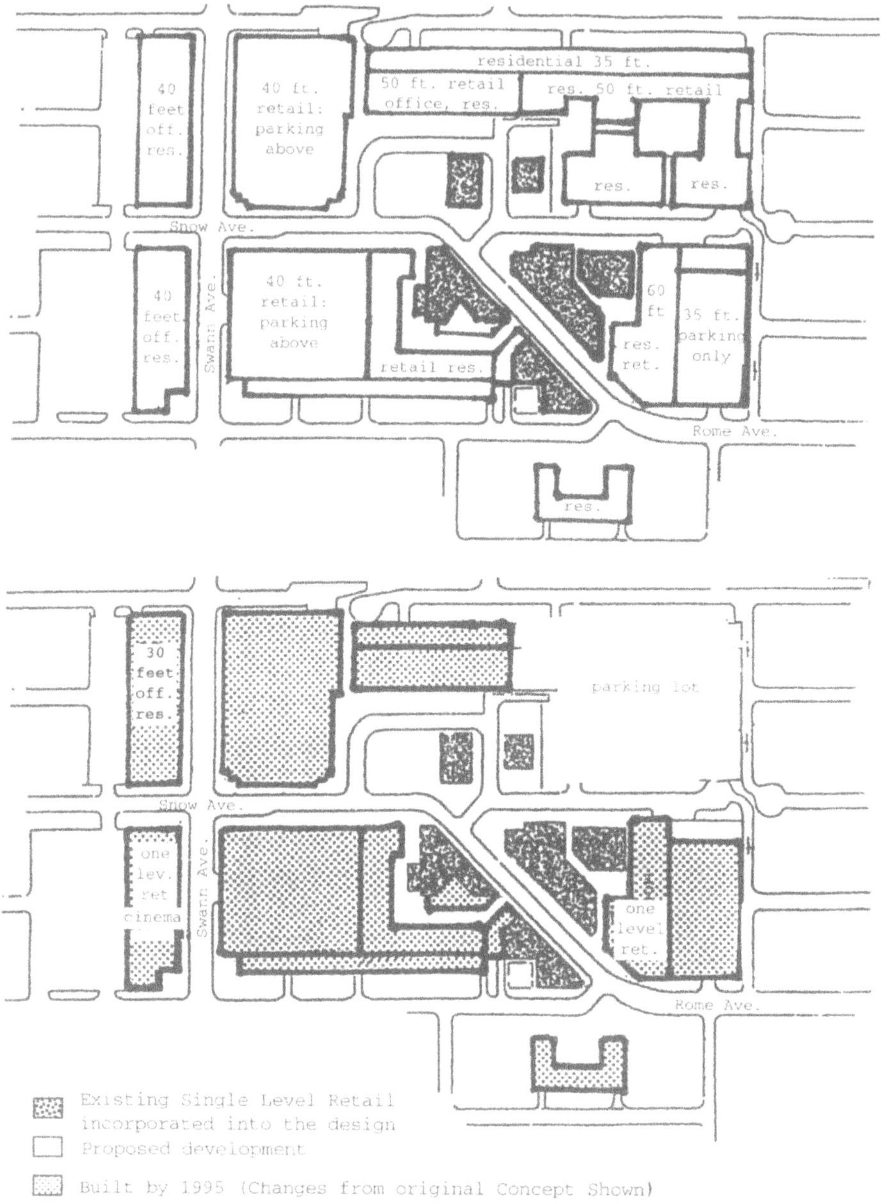

***Figure** 7. Old Hyde Park Village Initial Proposal, 1979, and Development, 1991.*

Amlea presented their plans to the Tampa City Council on July 19, 1979. The Amlea team, resplendent in expensive suits, polished, and articulate in speech, presented a stunning visual presentation, the likes of which had never been seen in Tampa according to several observers who were long-time area residents. But the Hyde Park residents, who were themselves articulate, well-dressed and professionals—architects, engineers, accountants, lawyers, and business executives—were not taken in by Amlea's high-tech graphics. Resident after resident protested the proposed development as "too dense, too big, out of scale, and out of sync" with Hyde Park's character. Many who had participated in the planning process that resulted in the Hyde Park Plan referred to that document, stating, "We already have a plan; why are you imposing this thing on us?" When one of the Amlea team members stated that plans should be flexible, one resident fired back "Flexible, hell...we just passed this thing three years ago and we looked at it again just a couple of weeks ago when we approved Tampa's version of Horizon 2000." More measured responses to the notion of "flexibility" included remarks to the effect that conditions in Hyde Park had not significantly changed in the three years following plan adoption to warrant rethinking of the "design commercial" designation for the site Amlea now owned.[8]

As the meeting concluded, the surprised Amlea representative realized that instead of dealing with a bunch of ignorant hillbillies they had encountered a poised, knowledgeable, and well-organized group of adversaries. Currying favor with elected officials behind the scenes wouldn't work, because the elected officials were not about to ignore a vocal middle class constituency. The Tampa City Council members remembered the stink Brandon residents had put up one year earlier when faced with a new landfill. Amlea would have to win the hearts and minds of local residents — at least enough to make a difference in the votes of a majority of HCPC and Tampa City Council members.

The Struggle for Plan Amendment Approval

The Amlea group was not only surprised by the adverse community reaction to Old Hyde Park, but disappointed as well. They had expected to breeze through plan amendment and rezoning procedures during the summer and fall of 1979 and, with all approvals in place, arrange for financing commitments before the year's end. Interest rates were beginning to rise by mid-1979 and the financial community nationally and internationally expected continued escalation of interest rates in order to counter inflation. But given citizen response at the July 19th City Council meeting, Amlea knew they must revise their plans then try again.

Meanwhile, the Hyde Park Civic Association, led by Lyn Whitelaw, its president, began to publicize this issue to the neighborhood residents, warning of its dangers. The publicity was spread by the association's monthly newsletter and flyers. By the end of 1979, Hyde Park residents were eagerly anticipating the revised Amlea proposal. The new proposal was unveiled to Hyde Park Civic Association members and the public at a city council meeting held on December 4, 1979.[9] The project had shrunk to 298,000 square feet of commercial space, but the office space had grown to 80,000. Instead of municipal parking lots with meters, private garages with free validated parking were substituted. The condominiums, now increased from 100 to 140, were placed on the development's periphery to be used as a buffer between the Old Hyde Park commercial and office uses and the surrounding residential blocks. Building heights were slightly scaled down to give the development a better fit with the neighborhood. Joe Bogdan, the architect for Amlea, promised that all shops would be of the highest quality, stating that the developers already had verbal commitment from Benetton, Ann Taylor, Banana Republic, and a new "yuppie" oriented chain, The Sharper Image. "There'll be no K-Marts here" said Bogdan in his charming manner. "Just give us a chance; you'll be proud of us." "We just want to be a good neighbor." Civic Association members listened politely, then told Bogdan that they were intrigued by Amlea's mixed use–pedestrian oriented approach and would be even more interested if it weren't so dense.

By February 1980 the recession had begun in earnest and the Swann–Snow intersection — now owned by Amlea, courtesy of James Burt — sprouted many vacant shops and office spaces. Amlea initiated a major public relations move by establishing a Tampa office in one of the vacant storefronts, which was remodeled with plush furnishings and a scaled drawing of Old Hyde Park prominently displayed in the window. Amlea also began to acquire several residential properties adjacent to the Swann–Snow commercial district, making the former owners very happy by paying top dollar in a market depressed due to the recession. Fresh from purchasing some much needed goodwill, Amlea went back to the City Council with a revised version of the site plan. While no major changes were made in terms of square footage or numbers of condominiums, the site was spread out from 6 to 12 acres. At the Council meeting held on March 12, 1980, residents again protested, calling Amlea's proposal too dense and too big.

In April 1980, the Historic Tampa–Hillsborough County Preservation Board, heavily influenced by Hyde Park residents, asked HCPC to review Amlea's plans, claiming that Old Hyde Park was incompatible with the surrounding area and in violation of Horizon 2000. The newly hired planning director for HCPC agreed to work with the Amlea group to develop a compromise plan suitable to Hyde Park residents. Suffering from massive budget cuts, a loss of 18 of 40 staff positions, and a forced move from the courthouse to a dusty old office building, HCPC was at an all-time low. The HCPC director knew that to take on Amlea directly and try to force Horizon 2000 and the Hyde Park Plan down their throat would risk HCPC's abolishment, especially if a nasty and protracted battle would result in the Tampa City Council's endorsement of Amlea's plans. Therefore, HCPC was as eager to find common ground with Amlea as Amlea was to make peace with the Hyde Park residents, while getting approval for most, if not all, of their development proposal.

In June 1980, Amlea, after conferring with Hyde Park residents, HCPC staff, and individual members of the Tampa City Council, publicly offered a compromise plan. The commercial space was reduced from 298,000 to 220,000, and the office space from 80,000 to 45,000, but the number of condominiums was increased from 140 to 197. While residents still protested, Amlea firmly stated that it had made its "final offer."[10]

Amlea then turned on the publicity. Mary E. Estes, a Tampa native and a former staff writer for the *Charlotte North Carolina News*, was hired as "communications director" for Amlea. Her initial task was to publish a handsome four-page monthly newsletter called "Old Hyde Park News." The first issue published, July 1980 (Vol. 1 No. 1), was headlined "Final Village Plans Show Reduction." The article went on to state:

> Completing our plans to redevelop Old Hyde Park has been a long time coming. We were anxious to create a plan compatible with what you, our neighbors, wanted to see happen in Hyde Park.
>
> So we came to you, and at a series of 29 public meetings last winter you told us your concerns. The size of the commercial component in our preliminary plans was mentioned as your primary concern. You also said you wanted the residential character of Hyde Park to be retained.

> Traffic was on your minds. What would happen to the neighborhood streets? And what about preservation? Would we save the homes? The trees? You asked about drainage, too.
>
> First, let us look at our completed plan and the changes we made as a result of your comments. Then, in this and in subsequent stories, we'll talk about Amlea's plans for preservation, traffic control and drainage.
>
> Our project, called Old Hyde Park Village, is modeled after the European concept of urban design. It mixes residential and commercial space around a central square.[11]

The article went on to talk about the plan's details, stressing that parking would now be free. The article also noted that a complete copy of Amlea's proposal would be available for review by any citizen at the local office.

Meanwhile, Hyde Park citizens opposed to Amlea's project began their own lobbying. By June 1980, Hillsborough County Planning Commissioners, including myself, had begun to receive personal handwritten letters from Hyde Parkers, protesting Amlea's plans. First, the letters were 1 or 2 per week, then 1 daily, then by September we were receiving 10–15 letters each day. The letters certainly got our attention, and we made a special effort to research the background of Old Hyde Park Village, the Hyde Park Plan, and the history of Hyde Park itself.

By October, the controversy was really heating up. Back in August, Amlea had requested a plan amendment for the site, which was a precondition to obtaining a rezoning from C-1 "neighborhood commercial" to Planned Unit Development Commercial. In September, City of Tampa planners privately endorsed Amlea after being convinced that taxes would increase by up to $1 million per year on their properties after development was complete. On September 28, 1980, HCPC staff released its report. It recommended downscaling Amlea's commercial space from 220,000 to 100,000 square feet, while recommending no major changes in office footage or number of condominiums. HCPC had been regularly castigated in the local press after a local Florida legislative bill (Chapter 78-524) had stripped away its zoning responsibilities and the greater than majority local government override provisions. On October 9, 1980, Tom Franklin attacked HCPC in an editorial entitled "County Planners an Endangered Species." Chiding the staff for recommending downscaling of Amlea's project, and calling them "dinosaurs," he urged the planners to be more political, especially where it appeared that the City of Tampa and recently elected mayor Bob Martinez favored the project.[12]

During the period of August through October, Amlea's propaganda machine, the "Old Hyde Park News," moved relentlessly to convince residents that their plan was in the neighborhood's best interests. Amlea certainly wasn't playing by the rules of civility. When that corporation purchased James Burt's holdings in March 1979 they *knew* what the zoning was and what the Hyde Park Plan called for on that site. The major works on shopping center development all warned entrepreneurs not to purchase land unless proper zoning is already in place.[13] But Amlea chose to take a chance, purchase properties at low-end costs, then by the use of slick presentation techniques, bamboozle elected and appointed officials into accepting their proposals, amending existing plans, and changing zoning requirements.

In the August edition of "Old Hyde Park News," (Vol. 1 No. 2) the headlines read "The Truth About Traffic." With a benignly smiling Harold Vick, Amlea's traffic engineering consultant, looking down on the imaginary peasants of Hyde Park, the article went on to state that instead of the Institute of Traffic Engineers standard of 49 trips per 1000 square feet of shopping floor area, a standard of 40 trips per 1000 feet should be used based on *their* analysis of trips at Bal Harbour Shops in the Miami area, Worth Avenue in Palm Beach, and Oakbrook Square in Palm Beach Gardens. With this questionable analysis the article went on to state a bottom line conclusion that traffic wouldn't be a problem. What the article conveniently forgot to mention was that the Bal Harbour, Worth Avenue, and Oakbrook Square developments were all located at the intersection of two major arterials, whereas Old Hyde Park would be built at the crossing of a minor arterial (Swann) and a collector (Snow), which intersected at a 45° angle rather than the normal 90° angle.

The September issue of "Old Hyde Park News" was headlined "Good News About Drainage." The article began with "If you are one neighbor who disliked an earlier drainage plan...rest easy. Amlea has a more efficient way to transport the rainwater from Old Hyde Park Village." The article went on to show how the drainage would be handled internally in retention basins onsite rather than placing a drain down one of the major streets, which would have flowed into a major sewer already over capacity. Other articles described how Amlea's new design had been deemed compatible with the neighborhood by the Historic Tampa–Hillsborough County Preservation Board and that no Development of Regional Impact study was needed, as the commercial and office square footage and number of condominium units were well under the threshold. On October 7, 1980 the HCPC staff issued a second report criticizing Amlea for violating Urban Land Institute standards in proposing a development that was much too dense for the neighborhood.

The October issue of "Old Hyde Park News" blasted HCPC for their staff report. Headlined "Why A Negative Report," this article read as follows:

> We were surprised and disappointed by the report from the staff of the Hillsborough County City–County Planning Commission. The report recommended against changing Tampa's land use map to accommodate Old Hyde Park Village.
>
> In denying our request, however, the staff said it recognizes the Village has unique qualities. And therein lies the confusion.
>
> Old Hyde Park Village is a mixed-use development that combines housing, shops, and offices. There is little history of such projects in Tampa and Tampa's land use standards do not address such developments.
>
> On one hand, the staff has said the Village conforms to Tampa's plan because it increases housing densities through developing multiple-family dwellings. It applauds Amlea's use of design standards and careful site planning. On the other hand, there is confusion about the size and nature of the commercial component. The staff has no internal guidelines for determining what type of commercial facility is acceptable in Hyde Park.
>
> Amlea is presently working with the Commission staff to make some modifications that are acceptable. At the same time, we are defining and clarifying what guidelines must be used for the modifications.
>
> We feel it is very important to the future of Hyde Park that it be developed in a planned manner. Without a plan, the properties on our site are subject to piecemeal development. Conceivably, under current zoning, and with no land use change, much of Old Hyde Park Village could be developed into freestanding, six story apartments, strip commercial and a variety of office uses.
>
> But that's not Amlea's goal.
>
> We want a quality development...a development that protects the surrounding neighborhood by buffering it with quality townhomes...a development that is landscaped and designed to conform with Hyde Park's existing architecture

> ...A development that offers such services to you as a park, a drug store, a grocery store, yet combines these services with high quality shops that would work in harmony with the neighborhood uses...
>
> ...A development that tightly seals the commercial inside the site instead of opening it to the residential streets...a development that protects rather than harms its surroundings.
>
> We are entering a critical point in our approval process. The question before city officials is: Do you want planned or unplanned development in your neighborhood? We believe that proper planning will save Hyde Park from haphazard, uncontrolled growth....[14]

To say that the article was misleading would be an understatement. In the first place, whoever wrote the piece knew that the staff recommendation was based on Urban Land Institute standards and criteria, so they deftly dropped in the phrase "the staff has no internal guidelines for determining what type of commercial facility is acceptable in Hyde Park" to make HCPC's planners appear stupid and incompetent. They could have mentioned that the Hyde Park Plan was an "internal guideline," but this was too much to expect. Actually, the Hyde Park Plan did indicate that the type of development Amlea wanted could be accommodated in the neighborhood's eastern edge on parcels designated "High Density Multi Family–Office" but Amlea didn't buy that land; it was probably too expensive! The notion that the existing zoning could be developed into "free standing six-story apartments, strip commercial" was misleading, too. The zoning on the Amlea site was C-1, which required 35-foot setbacks and a three-story height limit. There was no way Amlea could build a profitable development of *any* kind with C-1 zoning given the fact that the corporation had already spent $8.5 million for the property and several hundred thousand for consultants, including architects, engineers, lawyers and, of course, "Old Hyde Park News."

With the decision on the staff report for Old Hyde Park Village placed on the HCPC's November 3, 1980 meeting, the maneuvering by HCPC staff and the Amlea group intensified. In a letter dated October 9, 1980 to Bob Ferguson, Vice President of Amlea, Inc., and sent to their Hyde Park office, HCPC staff made a five-point proposal to justify the upgrading of the Amlea site to medium-intensity residential on the land use plan with a maximum of 250 dwelling units. The proposal included: (1) deletion by Amlea of 22,000 square feet of commercial space on the project's periphery; (2) inclusion of significant onsite active recreational facilities; (3) documentation by Amlea that it would take specific action to mitigate negative impacts on the Hyde Park Historical District; (4) removal of at least two unsightly residential units on the project's border; and (5) assurance that Amlea would provide a buffer of single-family dwellings on the southern boundary of the project.[14]

This letter was copied to all members of the planning commission. I thought it was an excellent quid pro quo, except for one thing. The staff letter to Ferguson stated at the beginning: *"It will be recommended that the project should be considered a neighborhood serving facility if it contains a maximum of 199,000 square feet of gross leasable area in the retail sector and the uses be limited to neighborhood serving commercial uses."*[16] I was furious when I saw that because every planner knows that the accepted maximum for neighborhood shopping centers is 100,000 square feet, and secondly, there was no way the commercial uses could be "neighborhood serving" as plans called for Jacobson's as the anchor tenant supported by regional "one of a kind" shops such as Benetton and Ann Taylor. Amlea was an upscale specialty center, which would attract people from all over the Florida West Coast. Its "neighborhood" would stretch east to Lakeland, west to the gulf, north to Gainesville, and south to Sarasota.

I called the planning staff and, almost yelling on the telephone, said "What's going on? You know that 199,000 square feet isn't a neighborhood center." A staffer replied, "People will walk to these shops." I retorted, "Some people will walk to these shops, but those coming from Clearwater, Gainesville, and Lakeland won't. I know you folks are in a tough spot trying to negotiate with Amlea, but don't give away the store. They have too much invested to walk away. If we stick to planning standards they will come around."

In order to strengthen the staff's bargaining position, I sent a letter to my fellow commissioners, with a copy to staff, insisting on maintaining professional standards. The letter, written on October 17, 1980, read as follows:

> Dear Fellow Commissioners:
>
> As a professional urban planner with 19 years of experience in the field and charter membership in the American Institute of Certified Planners, I would like to share some thoughts with you concerning the Amlea development proposal.

> I fully support the staff contention that any development on that site should be on the "neighborhood" scale, oriented primarily to residents of Hyde Park and adjacent areas. However, I strongly disagree with the staff's contention expressed in the October 9, 1980, letter to Mr. Ferguson of Amlea that a development of 199,000 square feet of gross leasable area (GLA) would constitute a "neighborhood" center. *Internationally accepted planning standards state that the maximum size for a neighborhood shopping center is 100,000 square feet of GLA.* While this figure is not rigid, certainly 200,000 square feet is not acceptable for a neighborhood center. Specifically:
>
> 1. The standard of 100,000 GLA for a neighborhood center was developed initially by George Nez in *Standards for New Urban Development — The Denver Background*, and adopted by the Urban Land Institute in their classic volume, *The Community Builders Handbook* (p. 267) in 1968. This standard was reaffirmed by the Urban Land Institute in *Technical Bulletin #69,* published in 1973. Other planning standard works such as DeChiara and Koppleman's *Urban Planning and Design Criteria* (1975) and *Shopping Center Design* by George Lion (1976) utilize this standard. The Urban Land Institute's latest publication, *Shopping Center Development Guide* (1977), on page 7, states that a neighborhood shopping center should have a range of 30,000 to 100,000 GLA.
>
> 2. Some will contend that Old Hyde Park Village will be a part of an "urban" complex, one that shoppers will most likely walk to, bicycle to; therefore, a large center won't produce the same amount of traffic as a suburban-oriented one. This may be true, but the design program as detailed by Amlea calls for a "specialty" center — one that will attract affluent shoppers from all over the Tampa Bay area. In my opinion, these two factors cancel out each other. Therefore, we must continue to use the recognized standard. Attached are two exhibits showing the standard. Note the reference on Exhibit 2 to "specialty" centers.
>
> The evidence is clear that the maximum GLA for a neighborhood center is in the neighborhood of 100,000 square feet. I feel so strongly about this matter that if any professional planner attempts to justify a development with 200,000 square footage of commercial space as a neighborhood center, I will report this person to the board of the American Institute of Certified Planners for violation of professional ethics.

I heard nothing from the HCPC staff; but at the Planning Commission meeting held on October 28, 1980 the staff report distributed to all present had two changes: the recommendation for commercial square footage was 130,000 and that for offices had shrunk to 30,000. Checking my references, I found that in the Urban Land Institute's *Community Builders Handbook,* 1968, on page 267 it noted that in setting standards, the Urban Land Institute's staff had noticed in their survey of 278 locations nationally, that neighborhood centers observed ranged from 11,700 to 130,000 square feet of gross leasable area (GLA). I wasn't going to quarrel over a measly 30,000 square feet; after all, this was better than Amlea's latest "on the table" proposal of 220,000 or HCPC staff's old figure of 199,000. Amlea then agreed to reduce their commercial space to 199,000 but no further.

The November issue of "Old Hyde Park News" was released in late October — just one week before the HCPC vote. The headline read *"HPPI Endorses Old Hyde Park Village."* HPPI was Hyde Park Preservation, Inc., a group of residents who resided in Hyde Park's southeast quadrant, an area of more expensive, mostly single-family homes. As traffic would come to Amlea's development from the north and west, the southeast quadrant would not be invaded by would-be shoppers seeking parking. That area would also avoid drainage problems should Amlea's latest plans change due to excess costs. The HPPI board bought Amlea's line about disastrous consequences should they be forced to abandon Old Hyde Park Village, as they stated in their endorsement:

> If Amlea is forced to abandon the project, the property might well be developed in a piecemeal haphazard fashion, that would permit little input from the neighborhood. The next development proposal might be larger still or contain structures higher than four stories. In summary, we are not pretending this development will not have a significant effect on the neighborhood; it will.... We believe however, that its overall effect is likely to be positive.[17]

Amlea rushed to get the City of Tampa on its side. Tampa's mayor, Bob Martinez, privately favored the project because of its positive tax ratables, but crafty politician as he was, he withheld endorsement until he could wring out every possible concession from Amlea. By the end of October, Amlea agreed to pay $525,000 for road improvements, $25,000 for sewer upgrades, $66,000 for new water mains, and $400,000 for drainage. Amlea's contribution totaled just over $1 million, making their total investment in excess of $10 million. The $1 million contribution to the

City of Tampa for infrastructure improvements amounted to a voluntary impact fee. After this agreement was hammered out, the City of Tampa's Bureau of Planning announced on November 2, 1980 their support for Old Hyde Park Village. But Mayor Martinez did not endorse the project then. He was waiting for the recommendation from the Hillsborough County Planning Commission.

The Hillsborough County Planning Commission Votes

There were several action items for the regular HCPC meeting of Monday, November 3, 1980, but only one was really important — the vote on our staff recommendation concerning Old Hyde Park Village. In the week before the meeting I received — as did my fellow commissioners — over 100 letters from citizens in Hyde Park with about 80% urging a yes vote on the staff recommendation — which was essentially a no vote for Amlea. The battle lines were now clearly drawn. The staff recommendation called for 130,000 square feet of commercial, 30,000 for office, and 256 condominium units. At the last minute the staff had decided to increase the condo numbers from 225 to 256 in exchange for 70,000 feet of commercial wanted by Amlea because there was a lower net traffic impact. Amlea's proposal, quite a bit different than the one first announced in July 1979, called for 199,000 square feet of commercial, 32,000 for office, and 225 condominiums.

The planning commission consisted of 10 members. Gordan Brunhild and I were professors at the University of South Florida; Fritz Rawls, Jr., Henry Brown, and Manuel Fernandez all owned their own businesses; Joe Chillura, Jr. was an architect in a large Tampa firm with a diversified practice; Barbara Myres was an urban planner with St. Joseph's Hospital; Robert Edwards, who represented Plant City, was an attorney; Wilbert Williams was a local realtor; and Warren Johnson, who represented the small city of Temple Terrace, was a professor at the local community college and an assistant basketball coach there as well.

The meeting began at 1:00 P.M. sharp in the cramped, hot conference room, which was warmer than usual as over 100 people jammed into a space that only seated about 25–30 individuals. Ron Ferguson and Joe Bogdan made Amlea's presentation. They stuck to the basics, as by now their design and the issue in general had been well publicized in the local press and even local television. Ferguson went over plan details, emphasizing that Old Hyde Park Village would cost $50 million to build and that Amlea had already invested or committed $10 million into the project, including over $1 million for public improvements. "We just want to be a good neighbor," said Ferguson in closing. HCPC then presented the staff alternative. The staff praised Amlea for their attention to architectural detail and innovation in urban design, but emphasized that the project as planned by Amlea was simply too big for the site.

After the presentations, representatives from the City of Tampa Planning Office and neighborhood residents spoke on the merits of Amlea's plan and the staff recommendation. Most neighborhood residents backed the staff plan, though Jim Landis of Hyde Park Preservation backed Amlea. By 5:00 P.M. it was time for the commissioners to speak. When my turn came, I said, "When everything is taken into consideration, the final issue is do we as commissioners back a report our staff has spent literally thousands of hours on or is the weight of evidence so strongly in Amlea's favor that we go against our staff and endorse their plan. I feel the burden of proof should be on Amlea, not our staff. Amlea's proposal is certainly innovative and the concept of mixed use is one that especially, with energy conservation as an issue, will be seen in new developments built between now and at least the end of this century. But Amlea has not shown enough for me to turn my back on our staff and the Hyde Park Plan. I am voting for the staff report."

When the vote was taken at 5:30 P.M., Barbara Myres, Wilbert Williams, Robert Edwards, Henry Brown, and I voted for the staff report. The other five commissioners voted for Amlea's proposal. With a 5–5 tie vote the rules called for no recommendation to be forwarded to the Tampa City Council. The next day Tom Scherberger wrote in the *Tampa Times* that both sides claimed victory — "Neither side had won but they hadn't lost either."[18] The real message to the Tampa City Council was that it was now a dead heat between Amlea's supporters and opponents.

On to the City Council

The City Council's vote was scheduled for December 16. This meant that each side had about five weeks to win friends and influence people among the City Council members. Amlea cranked up its propaganda apparatus once more. The December edition of "Old Hyde Park News" was limited to only two pages and used a soft sell approach. The headline was "A Summary of Our Village Plans" and the article quietly discussed plan details. Mayor Bob Martinez announced approval of Old Hyde Park on December 11, 1980, and his endorsement was noted on the newsletter but placed under the lead article.

The Hyde Park Civic Association wasn't standing still either. In their December 1980 newsletter, the headline read "Hyde Park Civic Association Stands Firm on Amlea Issue." The article stated that HPCA had *not* endorsed Amlea and the organization should not be confused with Hyde Park Preservation, Inc., which they characterized as "representing some individuals in the east of Rome, south of Swann Section of Hyde Park." HPCA claimed that they had more members in that area than did Hyde Park Preservation, Inc.

The newsletter attacked head on what it called the Amlea Myths, which were: (1) if Amlea doesn't get all it wants, it will pull out of Hyde Park; (2) if Amlea leaves, its land would be attractive to K-Mart, public housing, etc.; (3) existing zoning would allow horrible alternatives to Amlea; (4) there will be no serious impacts; and (5) Amlea's project will be a small quaint "village." HPCA's response to those "myths" were (1) Amlea has invested too much to pull out; (2) the land is too valuable for K-Mart or strip commercial, and with Reagan's election there won't be any new public housing; (3) existing zoning is so restrictive Amlea couldn't possibly make a profit; that's why they needed a plan amendment and a zone change; (4) if there would be no serious impacts, the HCPC planning staff would have endorsed Amlea's plans, and the two professional planners on the Commission (myself and Barbara Myres) voted against Amlea and for the planning staff; and (5) three 4-story garages does not make for a small quaint village. The newsletter closed with an announcement.

> *The most spirited event of the Christmas season!*
>
> "The public hearing on the Amlea project is scheduled for 7:00 P.M., December 16th, at Curtis Hixon Hall. Mark your calendar now for what promises to be an action packed shoot-out!!!"

Actually, the feisty Hyde Park Civic Association had its work cut out for them. The City of Tampa's elected officials all tended to favor the project because of Amlea's announcement that Old Hyde Park Village would generate up to an additional $1 million annually in tax revenue. Elected officials were also impressed that Amlea was willing to fork over a million dollars to upgrade water, sewer, and roadway infrastructure on its site. Tampa, by the late 1970s, was a city that was 100% built-up and now aging, with significant areas of blight and decline. Any boost in tax revenue was welcome, and it would be extremely difficult to get elected officials to take a look at the big picture and a long-range view.

The one thing Hyde Park residents who were opposed to Amlea's proposal had going for them was the Hyde Park Plan. I called councilwoman Sandy Freedman the day after our Commission meeting on Amlea to enlist her support in downsizing Amlea's plans. Councilwoman Freedman, who later served as Tampa's mayor (1987–1994), was a progressive, innovative official who had come to campus on several occasions to speak to my urban politics class. I said to her, "Sandy, the people in Hyde Park worked with our staff to create the Hyde Park Plan. Amlea comes in three years later and wants to ignore the plan so they can maximize profits. If they get away with this, no neighborhood in Tampa or Hillsborough County will feel comfortable working with public elected officials and their planning staffs." A person with a strong track record of supporting planned growth, Councilwoman Freedman replied, "Bob, that's something to think about. In meeting with the Amlea people tomorrow, I'll do what I can."

On November 19th Amlea announced that it was reducing its commercial space from 199,000 to 185,000 square feet. However, it increased the office square footage from 40,000 to 56,000. Our staff transportation planner calculated that this revision slightly reduced the number of trips. Our land use planners felt that the revision was significant because they doubted that a market for 56,000 square feet of office space existed. Hyde Park Civic Association spokesperson Lyn Whitelaw stated, "Amlea did that to make Sandy (Councilwoman Freedman) feel better."[19] I felt that perhaps one final letter to the editor of the *Tampa Tribune,* the city's leading newspaper, might help to modify Amlea's plans. So on December 9, 1980, I wrote to state my position on Amlea. The letter, published on December 12th, read as follows:

> Dear Editors:
>
> I was one of the five members of the Hillsborough County Planning Commission that voted against the "Old Hyde Park Village" pro-

posal presented by Amlea, Inc. While the Commission vote ended in a tie and no formal recommendation was sent to the Tampa City Council, I feel that it is my duty to communicate my sentiments on this matter to the City Council and the public, in general.

As a professional urban planner, I found the Amlea proposal to be a very imaginative, attractive, and forward-looking concept, even though it would create serious neighborhood impacts. Our planning staff felt the same way, and in a spirit of cooperation and professionalism, presented five alternatives to Amlea, all of which were rejected by this developer. While Amlea proposed 200,000 square feet for commercial use, 32,000 for offices, and 225 residential units, I preferred the staff's alternative that called for 130,000 square feet of commercial, 30,000 square feet for offices, and 256 residential units for the following reasons.

First, the Amlea proposal will create serious traffic, drainage, and aesthetic impacts. For example, according to Amlea's own studies, traffic will almost double on Swann between Howard and Rome as a result of their development. The street improvements proposed by the developer and the City's Traffic Department, addition of turn lanes, more traffic signals, and so-called intersection improvements, will certainly help but in no way are sufficient enough to carry the increased load at the present level of service.

Regardless of technical considerations, the most important factor is that over the past seven years, Hyde Park residents have been working with our Commission to prepare plans for their neighborhood. This effort produced the Hyde Park Plan, approved by the Commission and later incorporated into the City of Tampa's Horizon 2000 plan adopted by the City Council on June 28, 1979. Recognizing the residential character of Hyde Park, both plans called for a much lower intensity of commercial and office use than proposed by Amlea. On the other hand, both the Hyde Park Plan and Horizon 2000 favored development of the type suggested by Amlea in the area bounded by Platt, Boulevard, Swann, and Bayshore, the eastern end of Hyde Park.

Both plans were in force when Amlea began active land assembly in the Hyde Park area. Rather than complying with plans approved by elected officials, these developers are trying to change the plans for their own benefit using a modern day medicine-show approach, replete with colored slides, illustrations, and hired "experts." The irony of this whole case is that Amlea could never get away with these tactics in their orderly and well-planned home town of Toronto.

I sincerely hope that the Tampa City Council will work closely with Amlea to insure that the goals, objectives, and policies of the Hyde Park Plan and Tampa-Horizon 2000 are followed. To do any less will compromise the effort being made by all of us toward planned growth in Tampa and Hillsborough County.[20]

The city council public hearing was held on December 16, 1980, as scheduled. Over 500 citizens turned out in Curtis Hixon Hall, Tampa's major convention center. Amlea made what was well-known by that time as the "standard presentation," with illustrations and a small-scale model. Then Amlea unleashed their "secret weapon": a presentation by Charles Haar of Harvard University, and the nation's top expert in land use planning. Haar stated that planning must be flexible and endorsed Old Hyde Park Village. The City Council then heard residents testify for and against the Amlea proposal for almost three hours. Representatives from the Hyde Park Civic Association who were against the project were equally matched by those from Hyde Park Preservation, Inc., who supported Amlea. The Hyde Park Civic Association group was clearly tired after fighting for almost two years against Amlea, and with Haar's presentation they were at a loss for words. Finally, at 11:30 P.M. the City Council voted. All were in favor, and the next day, the *Tampa Tribune* reported that Amlea had won, but with major concessions having been made to area residents.[21]

By March 1981 all approvals were in, but because of high interest rates Amlea could not obtain suitable financing until early 1982. Construction began in 1983, the first phase opened in 1986, and by 1990, Old Hyde Park Village was substantially complete.

The Downsizing of Amlea

Economics did what the Hyde Park Plan, the Hyde Park Civic Association, and the Hillsborough County Planning Commission staff could not accomplish. Amlea's final plans, approved by the Tampa City Council, called for 185,000 square feet of commercial space, 56,000 square feet of office space, and 220 condominiums. By early 1992, 225,000 square feet of commercial space, but only 20,000 square feet of office space, and only 38 condominium units had been constructed. With the economy in recession by late 1991, no additional construction was forecasted for the remainder of this century.[22]

Old Hyde Park Village turned out to be wonderful. Because it was downsized, it enhanced the Hyde Park neighborhood. Traffic is not a major problem except during the Christmas holidays, and the shops and restaurants offer an ambience that one would have to travel to Atlanta or New Orleans to find. Compared to the original proposal, only 57% of the initial square footage was actually built, and the traffic impact is only two-thirds as great (see Table 4).

But if Amlea had their way with Hyde Park, the development would not have turned out so nice. The vacancy rate — now virtually zero — would have been very high, and "lower end" not-so-special stores such as Target, K-Mart, or Service Merchandise would have been lured in to fill the gap. If 225 condominiums had been built, as Amlea had originally planned, many would have been sold at auction.[23] Old Hyde Park Village works because so many people cared enough to fight Amlea and finally force that developer to compromise. As a result, everyone in the Tampa Bay area is better off because of the struggle.

Transactive Planning, The Middle-Range Bridge "Developeritis," and Professional Standards

Until the early 1970s there were basically three types of operational planning theories: (1) rational–comprehensive, (2) incrementalism, and (3) advocacy.[24] In 1973, John Freidmann published *"Retracking America: A Theory of Transactive Planning."*[25] Reaffirmed in a later book published in 1987,[28] Freidmann coined the term *Transactive Planning.* He defines transactive planning as:

> A style applicable to both allocative and innovative planning in which the processes of mutual learning are closely integrated with an organized capacity and willingness to act.[27]

We could accept Freidmann's definitions of *allocative planning* as "concerned with actions that affect the distribution of limited resources among competing users," *innovative planning* as "concerned with actions that produce structural changes in the guidance system of society," and *mutual learning* as "a process in which the processed knowledge of the planning expert is related to the personal knowledge of his (her) client in the joint exploration of problems and possible solutions to them." If we do so, then the evolution of Old Hyde Park Village represents a good example

Table 4. The Downsizing of Old Hyde Park Village.

Month, Year	Commercial Sq. Footage	Office Sq. Footage	Number of Condos	Total Sq. Footage[a]	Total No.[b] Trip Ends per 24 hr	Event
July 1979	350,000	40,000	100	510,000	18,585	Initial proposal to Tampa City Council
Dec. 1979	298,000	80,000	140	546,000	16,850	Concession to residents
June 1980	220,000	45,000	197	501,400	12,899	Concession to HCPC and residents: Avoid development of regional impact reviews
Oct. 1980	199,000	32,000	225	501,000	11,874	Additional concession to residents: Attempt to gain HCPC approval
Dec. 1980	185,000	56,000	220	505,000	11,482	Obtain approval from Councilwoman Sandy Freedman, other members of Tampa City Council
Dec. 1990	225,000	20,000	38	290,600	11,684	Project completed

[a] Average residential square footage per unit is 1,200.

[b] Based on 49.5 trips per 1,000 square foot of commercial space, 14 trips per 1,000 square foot of office space, and 7 trips per dwelling unit.

Sources: Wolfgang Homberger, *Transportation and Traffic Engineering Handbook,* Prentice-Hall (New York), 1982, and Frank So (Ed.) *The Practice of Local Government Planning* (International City Managers Association), 1979.

of transactive planning. Freidmann's visual model of transactive planning (see Figure 8) shows planners bringing to the table concepts, theory, analysis, processed knowledge (data compilation and analysis), new perspectives, and systematic search procedures, while the clients contribute intimate knowledge, realistic alternatives, *norms,* priorities, feasibility judgments, and operational details. If we view the planners as HCPC, and City of Tampa staff and Amlea along with the residents as embodied by the Hyde Park Civic Association as "clients," we find that both sets of participants had the requisite for transactive planning, namely, specifics to contribute, an organized capacity for processing knowledge, and a willingness to act. Amlea certainly knew what they wanted in terms of a design concept and a specific development program. HCPC staff and those of the City of Tampa definitely understood that the key to their success lay with balancing Amlea's plans with neighborhood concerns assuring an optimal fit between that project and the surrounding environment. By mutual interaction, both sides gave in some, and the result was a plan that everyone, including the neighborhood residents, could live with.

The role of the Hyde Park Civic Association is interesting. One could view this group as advocates for Hyde Park and an equal player in the transactive planning game.[28] Certainly this group had resources to bring to the table, namely, a large number of articulate residents opposed to Amlea's plans. HPCA also had an advantage that is usually not available to advocacy groups. They were like Amlea staff: white, upper middle class, and well educated. HPCA and Amlea staff had attended the same type of professional schools, shared the same values, dressed alike, and acted alike. The sparring between Amlea and HPCA had the overtones of a Harvard–Yale football game or a Pan-Hellenic contest between two rival college fraternities. This author has to speculate that if HPCA represented poor, minority residents, Amlea might not have been so willing to bargain and compromise.

But HPCA failed to provide one crucial element — *an alternative plan to Amlea's proposal.* This group had "in-house" resources of planners, architects, landscape architects, and civil engineers. Without a tremendous amount of effort, HPCA could have — between December 1979, when it was clear that Amlea wasn't going to disappear, and November 1980 — developed their own plan for the Swann–Snow commercial district. The plan could have been in schematic design and perspective form

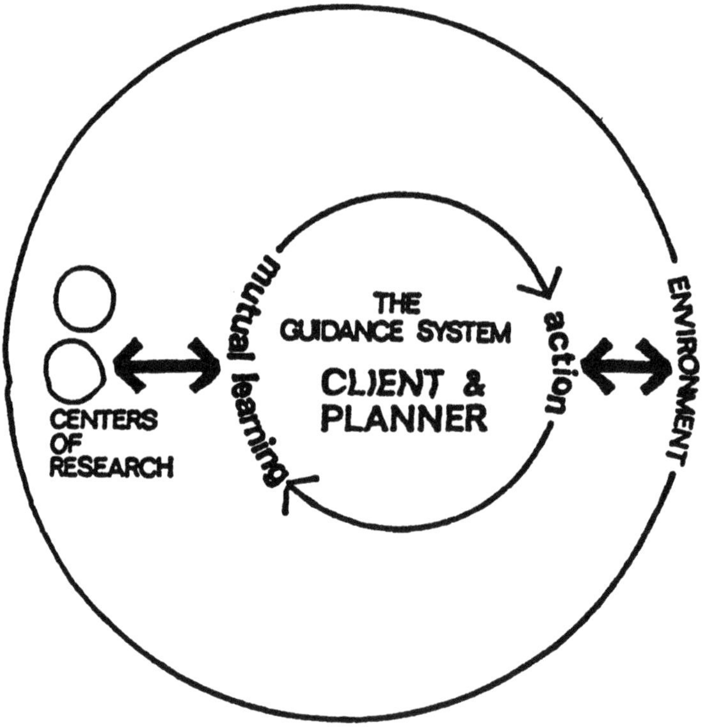

PLANNERS CONTRIBUTE	CLIENTS CONTRIBUTE
• concepts	• intimate knowledge of context
• theory	• realistic alternatives
• analysis	• norms
• processed knowledge	• priorities
• new perspectives	• feasibility judgements
• systematic search procedures	• operational details

Figure 8. *A model of transactive planning. Source: John Friedmann,* Retracking America: A Theory of Transactive Planning. *Anchor Press/Doubleday (Garden City, New York, 1973), page 187, reprinted with permission.*

along with a design program showing allocation of retail commercial, office, and residential space. Such a plan would have raised the ante and incrementally forced Amlea to either concede more space or make larger contributions to infrastructure provision. But HPCA chose to behave negatively; they opposed Amlea, but offered no alternative. Finally, when Hyde Park Preservation, Inc. endorsed Amlea's proposal — a *positive* action — HPCA was left holding a bag of negatives, and lost.

Also, the final project was reduced in terms of building heights, and the final design was much more compatible to the neighborhood than the original. In short, transactive planning worked in this case.

Table 4 shows just how well the transactive planning process worked to downsize Old Hyde Park Village. The initial proposal called for 510,000 total square footage with a traffic impact of 18,585 trip ends per 24 hours. Amlea's redesign in June 1980, while maintaining the same square footage, reduced traffic by one-third. The final design presented to the Tampa City Council reduced trips by another 1,500 vehicles and the build-out saw the square footage almost cut in half.

The vehicle that set the stage for a transactive planning relationship between the major parties was the Hyde Park Plan. Horizon 2000, passed in late 1977, was a long-range comprehensive plan with a time frame of almost 25 years. This plan showed the Hyde Park area as "urban development" and the Swann–Snow area as an "urban high density node." Had there not been a Hyde Park Plan, Jim Burt — and later Amlea — could have justifiably claimed that Horizon 2000 endorsed their 350,000 square foot shopping center. But the Hyde Park Plan, being site-specific and a 5–10 year *development* plan for the neighborhood, defined the character of the "urban high density node" as one of design commercial (i.e., a neighborhood shopping center). Even though HPCA did not present an alternative design scheme to Old Hyde Park Village, just the presence of the Hyde Park Plan created the climate for negotiation. The Hyde Park Plan became a neighborhood safeguard because it said what was specifically desired for the site in question.

While it is obvious that mid-range plans can further assist both public planners and developers to make mutually beneficial land use allocation decisions, it is interesting to note that neither the 1975 nor the 1985 Local Government Comprehensive Planning Acts mandate a mid-range effort. The 1993 revisions adopted by the Florida legislature also fail to recommend mid range planning. HCPC was moving in the direction of mid-range planning with its "sector" plans in early 1978, but 1979 and 1980 budget cuts made sector planning impossible. This author is of the opinion that, given the eventual conflict between government agency planners and developers, a mid-range plan is a must in order to satisfactorily resolve this conflict.

"Developeritis"

This author chooses to define "developeritis" as a dreaded disease whereby infected persons willfully ignore long-range policy plans and mid-range development plans in order to maximize profits. It is caused by greed, callous neglect for fellow human beings, and a superiority complex that in turn causes carriers of the virus to feel that they are better than neighborhood residents and government workers who, if they were any good, would be employed in the private sector.

We saw in this case how developeritis works. Knowing what was called for in the Hyde Park Plan first passed by the Tampa City Council in June 1976 and reaffirmed by this same body in June 1979, Amlea purchased the property anyway and then had the nerve to propose on July 2, 1979 a project that was 180 degrees away from the "design commercial" called for by the Hyde Park Plan. As I mentioned earlier, the type of use Amlea wanted was appropriate on the neighborhood's eastern edge as recommended by the Hyde Park Plan. But one of the symptoms of developeritis is the urge to buy cheaper land, then demand that it be rezoned to meet the developer's needs, using the rationale that whatever's good for the developer is good for the neighborhood and the city as well.

There is no known cure for developeritis. However, this disease can be managed by local governments having available the full set of planning tools, including a long-range comprehensive plan, a mid-range development plans, a full set of development regulations, and an airtight zoning ordinance. Unfortunately, even with the 1975 and 1985 mandated planning legislation, too many of Florida's local governments lack, in a sharpened, ready state, some, if not all of these tools.

Maintaining Professional Standards

Having all of the previously mentioned tools is meaningless if professional planners bend over backward to appease either development interests or elected officials and by doing so violate their own professional standards. Transactive planning does not mean abandoning standards. As Freidmann says

> ...if the planner enters the realm of action, his word must be responsible. He is no longer setting forth alternatives (as under the old decision scheme) but advocating points of view that will affect the lives and well being of others. His word is that of a professional and his professional ethics must therefore include an injunction against irresponsibility. This is a question of consciousness that each planner must settle for himself. The results of societal action are manifold. But difficulty of measurement should not be taken as a convenient escape from personal responsibility.[29]

This case presents documented evidence that HCPC staff, in their eagerness to resolve conflict by appeasing Amlea, were prepared to abandon recognized standards for categorizing types of retail commercial districts. In my 35-year career as a professional planner, I have seen all too often examples, such as the one in this case, where planners have forsaken personal responsibility in order to gain acceptance and guest membership in "Club Developerville." What is lost by this action is the quality of life and well-being of others, as well as the planner's own self-respect.

Summary

In this case, our public planners — both those from HCPC and from the City of Tampa — seemed to do almost everything right. First, a neighborhood plan was prepared by the residents of Hyde Park working with HCPC staff. That plan was revisited and reaffirmed as a link to the Tampa version of the long-range Horizon 2000 Comprehensive Plan. When, less than one week later, Amlea presented a project that was in direct conflict with the Hyde Park Plan, neighborhood residents strongly resisted. Armed with the mid-range Hyde Park Plan, neighborhood residents, HCPC staff, and Amlea, the Toronto-based developer, entered into a transactive relationship between July 1979 and December 1980. The results were significant, as the design was altered to make a better fit with the neighborhood and the traffic impact was reduced by over one-third. Also as a direct result of the bargaining between Amlea and the City of Tampa planners, Amlea agreed to provide over $1 million for infrastructure improvements, something that was not part of their original proposal. Due to the tight money and recessionary circumstances of 1980–1983, Amlea changed the design once more and at build-out only 57% of the originally planned square footage was actually constructed though the traffic impact was virtually unchanged from the December 1980 proposal approved by the Tampa City Council.

One might speculate that economics alone determined what was eventually built. This author disagrees. Amlea fully intended to get all approvals *and* financing by late 1979, and then build a 350,000 square foot shopping center with 40–50 foot high buildings and municipal parking lots in the middle of a residential area with little or nothing in the way of a voluntary impact fee to mitigate disruption to the neighborhood. The shopping center would have come on line in 1982 at the height of the national recession, and Amlea would have resorted to filling the space with any and all types of tenants. The result would have been an ecological, environmental, social, and economic disaster both for Hyde Park and Tampa.

This scenario was never played out because of a number of factors. The most important reason was the presence of the Hyde Park Plan — that mid-range development document that went so much further than a comprehensive plan because *it specified exactly what type of retail commercial use was appropriate for the Old Hyde Park Village Site.* This plan set the stage for a process of transactive planning that was reinforced by the fact that mutual respect existed among all parties because all groups were essentially alike in terms of education, social class, and race. Again, this author wonders: if the residents were low income and minority, would either the developers or the public planners show the same level of respect for them, or would they join forces to give these "neighbors" the short end of the stick? Given historical precedent, especially the comparisons to the actions of governmental elites in the previous case, I think they would have done the latter.[30]

Regardless, the outcome in this case was a happy one for all sides. The greed of the 1980s resulted in a multitude of shopping centers, office parks, and so called "planned unit developments" financed and built under questionable forecasts and analyses, which even by the mid-1990s remain half empty and paying an extremely small percentage of anticipated tax revenue. Many of the institutions that underwrote their development are out of business and part of the one trillion dollar fiasco of failed financial institutions. This did not happen in the case of Old Hyde Park Village because, for the most part, public planners and neighborhood residents stood up to be counted. This time, the planners won, even though the process took years and the outcome was always in doubt.

Notes

1. Charlton W. Tebeau. *A History of Florida,* University of Miami Press (Miami, Florida), 1971, pp. 248-49.
2. Martin Meyerson. "The Middle Range Bridge for Comprehensive Planning," *Journal of The American Institute of Planners,* 1956.
3. Hillsborough County Planning Commission. *The Hyde Park Plan; Tampa Urban Core Study,* p. 3, par. 2, 1976.
4. Ibid., p. 3, par. 2.
5. Resolution Number 8950F. Tampa City Council, June 17, 1976.
6. Hyde Park Plan, op. cit. Note 3, p. 14.
7. J. Ross McKeever. "Shopping Center Zoning," Urban Land Institute Technical Bulletin No. 69 (Washington, DC, 1977).
8. Dana Kerbs. "Realtor Unveiled Plan for Hyde Park Area," *Tampa Tribune,* July 20, 1979, p. B1. The author attended this and all other public meetings on the Amlea proposal.
9. Gil Klein. "New Hyde Park Center Plans Presented," *Tampa Tribune,* December 5, 1979, p. B5.
10. Kevin Kalwary. "Business Space Reduced in New Hyde Park Plan," *Tampa Tribune,* June 14, 1980, p. B1.
11. Old Hyde Park News: Vol. 1, No. 1, July 1980, p. 1.
12. Robert Catlin. Letter to the Editor, *Tampa Times,* October 14, 1980, published October 28, 1980.
13. See for example *The Community Builders Handbooks,* Urban Land Institute, Washington, DC, 1968, J. Ross McKeever, "Shopping Center Zoning" in *Urban Land Institute Technical Bulletin 69,* Washington, DC, 1977 and George Lion, *Shopping Center Design,* Montreal, 1976.
14. Old Hyde Park News, Vol. 1, No. 4, October 1980, pp. 1, 2.
15. Hyde Park Recall: Hyde Park Civic Association, Inc., November 1980, pp. 1 and 2.
16. Letter from HCPC Executive Director to Bob Ferguson, Vice President, Amlea Inc., October 9, 1980.
17. Old Hyde Park News, Vol. 1, No. 5, November 1980, p. 1.
18. Tom Scherberger. "Commissioners Voted on Merits, not Principles," *Tampa Times,* p. 1A, 11-4-80.
19. Gil Klein. "Hyde Park Center Reduces Size Again," *Tampa Tribune,* November 20, 1980, p. B1.
20. Robert Catlin. "Don't Abandon Hyde Park Plan Goals," Letter to the Editor, *Tampa Tribune,* December 12, 1980.
21. Steve Piacante. "Amlea Wins Approval From City," *Tampa Tribune,* December 17, 1980, pp. A1, A8.

22. Between December 1980 and 1985, the development order was amended twice. The final development order increased the commercial footage by 40,000 over what had been agreed to by the Tampa City Council, but the office square footage was reduced by 36,000 and 182 condominium units were deleted. Amlea requested these changes due to market conditions during the 1980–1983 recession. Information as to the build-out: Source, Tia Faulkner, Marketing Director, Old Hyde Park Village.
23. For example, during the period of 1989–1991, there were three separate auctions to unload unsold condominiums at Donald Trump's luxury tower in West Palm Beach. Of 225 units, 20 still remain unsold. The promise of additional tax revenue to offset improvements to infrastructure installed by the city of West Palm Beach has not materialized. Source: Ms. Shirley Wray, former Community Development Director for the City of West Palm Beach.
24. Andreas Faludi. *A Reader in Planning Theory*, Pergamon Press (New York, London), 1973.
25. John Freidmann. *Retracking America: A Theory of Transactive Planning*, Anchor Press (Garden City, NY), 1973.
26. John Friedmann. *Planning in the Public Domain*, Princeton University Press (Princeton, NJ), 1987.
27. Friedmann. op. cit. Note 25, p. 247.
28. See Paul Davidoff. "Advocacy and Pluralism in Planning," Vol. 31, November 1965, *Journal of the American Institute of Planners.*
29. John Freidmann. "Notes on Societal Action," *Journal of the American Institute of Planners*, Vol. 35, No. 5, 1969, pp. 311–18.
30. This was the case for so many urban renewal projects across the nation during the period of 1950–1990. For instance, see Chester Hartman's *Yerba Buena: Land Grab and Community Resistance*, San Francisco: Glide, Memorial Press, 1974. In this work, Hartman examines the process whereby 3,000 units of housing occupied by retired merchant seamen, longshoremen, and other middle-aged workingmen were demolished for a convention center between 1970 and 1974. Leading the effort to uproot these residents without adequate relocation were San Francisco's mayor, the Board of Supervisors, the Department of City Planning, the San Francisco Urban Renewal Agency, building and trade unions, building contractors, and land developers. The residents formed a group called Tenants and Owners Opposed to Redevelopment (TOOR) and put up a spirited battle. However, the site was cleared by 1974, and even though 300 units of replacement housing were built by 1980, mainly as a result of the protests, many of the original TOOR members were dead. The Robert Moscone Convention Center finally opened in 1982.

6 The West Palm Beach City Center: A Succession of Plans

Introduction

In West Palm Beach, during the period of 1978–1989, no less than three development plans for the city's downtown core were prepared, paraded before government officials, private sector participants, and citizens, and then formally adopted. Each one was more grand, more ostentatious than the one that preceded it. Each successive plan resulted in less credibility than the one that preceded it. After the adoption of each plan, public sector planners and citizens became more and more alienated from the preparation/adoption process, each plan was amended with greater and greater ease, and by early 1992 few leaders in the city and metropolitan area's business community took planning very seriously.

This chapter examines not only each attempt at using planning as a catalyst to revive a moribund, albeit quaintly beautiful downtown, but even more, the process by which these plans were amended and redrawn. The changes were instigated not by the planners, but by development interests. The developers dreamed up bigger and bigger schemes to make money for themselves and then led the city's planners virtually by the nose to reshape the plans in their own image. The role (or lack of it) of the 1975 Local Government Comprehensive Planning Act (LGCPA) and the 1985 Local Government Comprehensive Planning and Land Development Regulation Act (LGCPLDRA) in the process will also be examined.

Background

The first permanent residents of Palm Beach County arrived during the 1870s and most settled on the east shore of Lake Worth on what is now the site of the town of Palm Beach. In 1892, Henry M. Flagler visited the area looking for a route to Miami for the extension of his railroad. Completely taken in by the area's beautiful scenery and climate, Flagler decided to create the resort community of Palm Beach and soon initiated construction of the Royal Poinciana Hotel.[1] Needing a site to initially house the construction workers for his railroad and hotel and eventually provide for the commercial needs of Palm Beach, Flagler purchased property from Captain O.S. Porter and Louis Hillhouse on the west shore of Lake Worth.[2] In November 1893, Flagler filed the original plan for the Town of West Palm Beach. The town extended from Lake Worth on the east to Clear Lake on the west (see Figure 9). Flagler's Florida East Coast Railroad reached West Palm Beach the same year, bringing building materials, tourists, workers, settlers, and of course, real estate speculators. The first lots in the town were sold in February 1894, the community voted to incorporate in November 1894, and the first census, taken the following year, showed that 1,192 persons resided in the city.[3] By 1900, West Palm Beach had a water pumping station, a sanitary and storm drain sewer system, electricity, paved streets, and a telephone system. West Palm Beach became a city in 1903, was designated as the County Seat, and by 1920 was well established, with a permanent population of 8,659. The completion of the Palm Beach Canal in 1917 linked West Palm Beach with the newly developing agricultural areas in the "Glades" region of Lake Okeechobee, and this city became a major shipping point for the Glades farm products during the early 1920s.

West Palm Beach's major period of growth came during the Florida land boom of the 1920s. Commercial activity was centered on Clematis Street, and major office and commercial structures were erected, some of which were 8, 10, and even 14 stories tall. Although by the late 1970s the Florida land boom went bust, the city's population rose to 26,610 by 1930, a whopping 207.3% increase over that in 1920, and 33,693 by 1940, which represented a 26.6% increase over the previous 10-year period — this, despite the Great Depression of the 1930s. World War II ushered in another boom as an Army Air Corps base was established 4 miles from downtown on what is now Palm Beach International Airport. By 1950, the city's population had increased to 43,162, a 28.1% jump over that in 1940. During the 1950s, downtown West Palm Beach prospered, as it always had since the early 1920s. Stores were filled with goods and customers, offices were fully occupied, and new government buildings — both

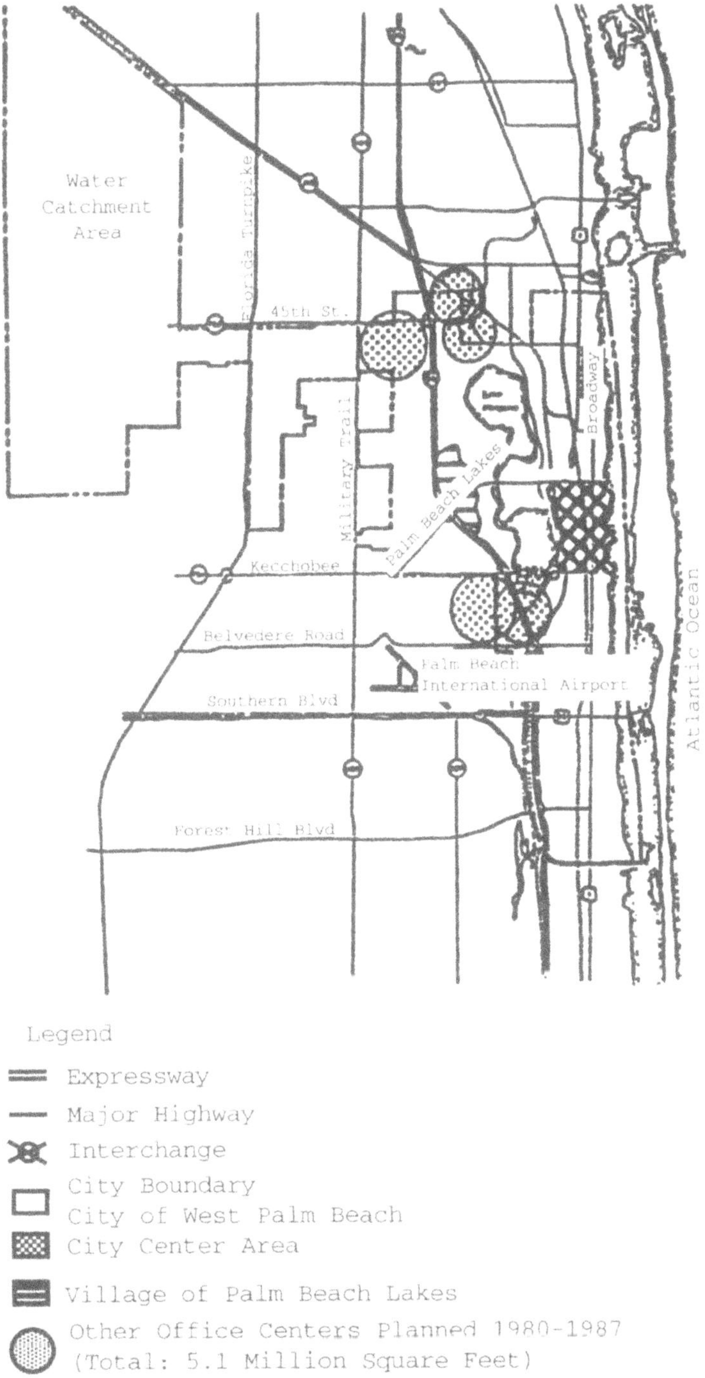

***Figure 9.** Vicinity sketch of West Palm Beach area.*

state and federal — were planned and eventually constructed. By 1960 West Palm Beach was virtually 100% built-up east of Australian Avenue, and total population virtually peaked at 56,208, 30.2% more than in 1950.

While business owners, realtors, and those tied by investments to the old downtown area were rubbing their hands in glee with the "cha-ching" of cash registers ringing in their ears, unanticipated changes were just about to occur. First of all, until the early 1950s, population in Palm Beach County was concentrated along the old coastal ridge communities of Jupiter, Riviera Beach, West Palm Beach, Lake Worth, Boynton Beach, and Delray Beach. Affluent and upper middle class whites settled between the western shore of Lake Worth and the Intercoastal Waterway on one side and U.S.-1 on the other. On both sides of U.S.-1 and going west to the Florida East Coast Railroad, strip commercial establishments dotted the road all the way south to Ft. Lauderdale and Miami. In West Palm Beach, between the Florida East Coast and the Seaboard Coast Line Railroad, blacks, poor whites, and industry were sandwiched together in the same linear pattern, and west of the Seaboard tracks the swamplands began. But in the late 1950s, the Florida turnpike was planned about 10 miles west of the Seaboard's line. The U.S. National Defense Highway Act of 1956 provided for the construction of 40,000 miles of controlled access interstate highways, and one of them, I-95, was planned to be located just west of the built-up portion of West Palm Beach. With I-95 connecting the Boston–Washington, DC megalopolis with Miami, tourists and commercial operatives would now have a new, fast, and direct route to connect these two areas, and this route just happened to run past West Palm Beach. I-95 and the Florida Turnpike now opened the adjacent portion of Palm Beach County for development.

In 1957, the Perini Land and Development Company purchased 4,200 acres of swampland from the City of West Palm Beach lying between Australian Avenue and I-95 just north of the Palm Beach International Airport. Lou Perini was the owner of the Boston Braves baseball team, which had for years trained in West Palm Beach. Although the Braves had moved to Milwaukee in 1953 and would move to Atlanta in 1966, they still held spring training in West Palm Beach. Perini, who like Flagler before him had fallen in love with Palm Beach County's weather and beauty, decided to make a longer-term commitment to the area. Between 1960 and 1972, Perini drained the swamps and constructed a development known as the Village of Palm Beach Lakes. Containing almost 1,500 acres, this development by 1995 consisted of 7,400 residential units, 100 acres of commercial properties, including 800,000 square feet of office space in 12 story towers, and 2 golf courses. The linchpin of the Village of Palm Beach Lakes is the mammoth Palm Beach Lakes Mall, a 1.4 million square foot shopping center right at the intersection of I-95 and Palm Beach Lakes Boulevard with five major department stores and a multi-screen cinema. It opened in 1967 and was expanded and remodeled in 1989. The City of West Palm Beach assisted Perini by constructing on his land the Palm Beach Convention Center and new training sites for the Braves and the Montreal Expos. Palm Beach Lakes Boulevard was widened to 6 lanes between downtown and I-95 in the early 1970s, enabling travelers to conveniently move between the mall and downtown.

When Perini's plans were made known, individuals and firms connected to downtown West Palm Beach became concerned. In 1963, the City of West Palm Beach hired well-known planning consultant Milo Smith to prepare a redevelopment plan for downtown. Renewal was sorely needed because downtown, although prosperous even in the early 1960s, was aging, as the majority of buildings were originally constructed in the 1920s, parking was woefully inadequate, and traffic jams were of troubling proportions. Milo Smith's plan was completed in 1965 and it was a state-of-the-art downtown redevelopment document calling for the usual facade treatments, pedestrian malls, and strategically located parking garages. Between 1965 and the early 1970s many merchants and property owners took heed of some of Smith's ideas and made at least cosmetic improvements. But after the Palm Beach Lakes Mall opened, all of the department stores and almost 50 other smaller establishments left downtown for the new surroundings, which were only three miles away but much closer to the new center of population, which was moving west toward the Florida Turnpike. By 1975, many of the downtown property owners who *couldn't* leave were hoping that the newly enacted Local Government Comprehensive Planning Act (LGCPA) would provide some renewed hope.

The City Center Plan of 1978: A Modest, Measured Effort in Downtown Revitalization

Unlike many other Florida communities of similar size, West Palm Beach didn't have to be dragged kicking and screaming into a planning effort upon passage of the 1975 LGCPA. In 1965, Milo Smith's Central Business District plan referred to earlier was actually part of a citywide long-range comprehensive plan that included neighborhood profiles, a capital improvements program, and revised development regulations.[4] Although Florida did not pass enabling legislation, allowing local governments to prepare and adopt comprehensive plans, until 1969, the Milo Smith document still served as an informal advising tool for decision making in zoning amendments and related public actions.

West Palm Beach's elected officials, business leaders, and citizen activists became concerned about possible stagnation of their city when the 1970 U.S. Census revealed a population of 57,575, only one thousand more than in 1960. On the other hand, the county's population had increased from 228,106 in 1960 to 348,993 ten years later. While the Village of Palm Beach Lakes was developing within the city limits of West Palm Beach, local planners forecasted a shift in business and residents from the central core initially to the city's fringes, but soon thereafter outside of the city itself into surrounding suburbs. Therefore, during the early 1970s, the city's planning department began to compile data from the U.S. Census and related sources in an effort to update the 1965 Milo Smith Plan. When LGCPA was passed, the planners only needed to reshape the format to comply with that legislation's requirements. While the 1975 LGCPA did not require the city to prepare mid-range development plans, West Palm Beach in fact did just that. The city was divided into eight planning "sub areas." Not only were land use plans prepared for each sub area but, where appropriate, development recommendations were made and then tied to the city's ongoing capital improvements program. One of these sub areas, the *City Center,* is the subject of this case study (see Figure 10).

The City Center sub area included not only the Central Business District, but also three district residential neighbors (see Figure 11). The entire sub area is bounded by Palm Beach Lakes Boulevard on the north, the intercoastal waterway/Lake Worth on the east, Okeechobee Boulevard on the south, and Australian Avenue and Clear Lake on the west. Just north of the CBD core is the Brelsford-Gruber-Carlberg neighborhood, containing high-value single-family and multi-family residential uses on the eastern end and commercial and industrial uses west of Dixie Highway, which is U.S. Route 1. This neighborhood was losing population, with commercial/office uses intruding on residential blocks. The plan-

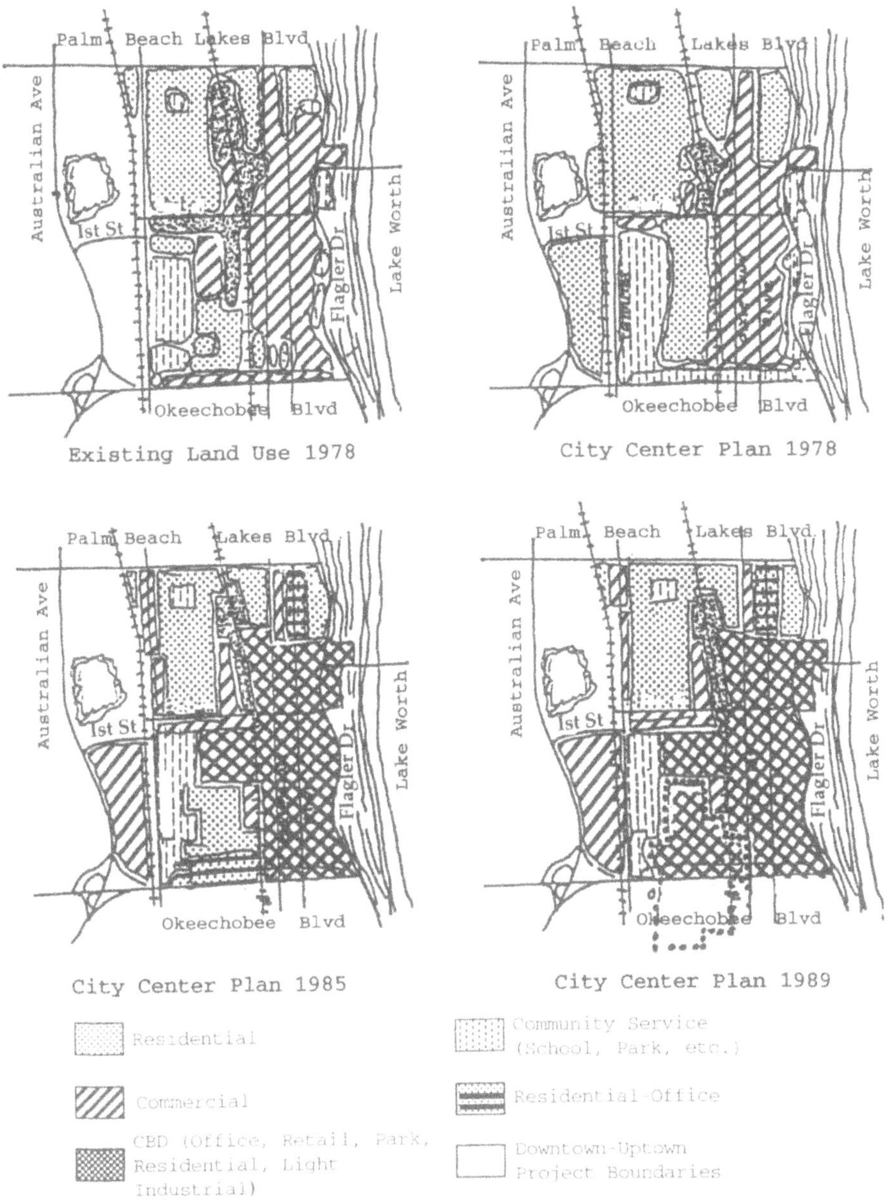

Figure 10. *A succession of plans: West Palm Beach City Center.*

ning staff realized that preserving and even enhancing residential use was vital to maintaining a 24-hour diversified "life" in the Central City. West of Brelsford-Gruber-Carlberg, just across the Florida East Coast Railway

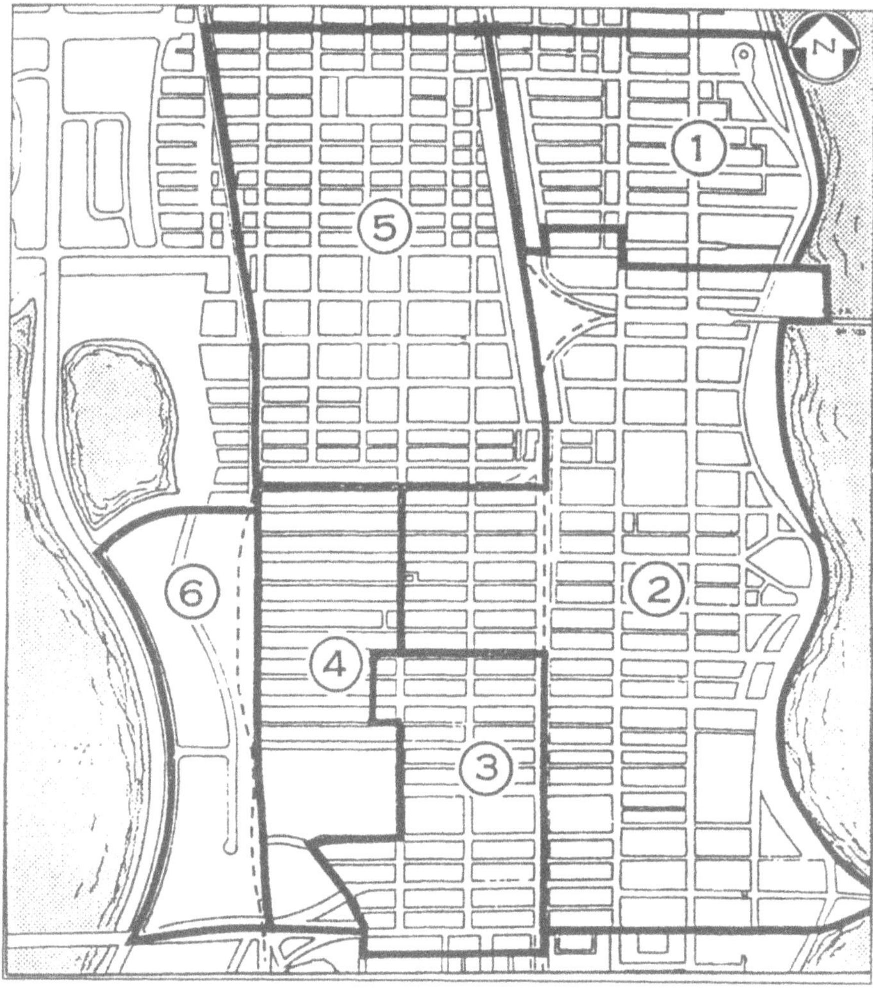

1. Brelsford-Gruber/Carlberg
2. Central Business District
3. Garden District
4. Government Hill
5. Northwest Neighborhood
6. Clear Lake

***Figure 11**. City Center sub areas. Source: City of West Palm Beach Department of Planning and Community Development. Reprinted with permission.*

tracks, was the Northwest Neighborhood. This neighborhood was the center of West Palm Beach's African American community and home of several large churches which drew their congregations from all over the county. The residential and commercial blocks had deteriorated over time, but there remained several well-kept residences, and the neighborhood was self-contained, as no east-west streets between Third and 11th crossed the tracks. Sitting on the coastal ridge, several blocks in Northwest had spectacular views of Lake Worth, Palm Beach, and the Atlantic Ocean itself. Also, the neighborhood elementary school, Palmview, was formerly the Colored Industrial School, founded in 1916 as the first high school for African-Americans in Palm Beach County. With many of the buildings constructed by students and teachers during the period of 1920–1940, Palmview was a community treasure and a major focal point. All of these assets outweighed the liabilities of deterioration, and Northwest lent itself to a variety of treatment programs, including rehabilitation, in-fill housing, and infrastructure upgrading. South of this neighborhood was an area named "Government Hill" by the planners. It contained Twin Lakes High School[5] and the County, Federal, and State Government Office center, which was expected to grow over the years as the county's population continued to increase. Just east of Government Hill was the "Garden District." Formerly a well-kept mixed single- and multi-family residential area, the Garden District had deteriorated badly by the early 1970s and needed revitalization.

The planners offered a modest sketch showing future and use for the City Center (see Figure 10). The area east of the Florida East Coast Railroad was generally reserved for commercial/office use, with the exception of a residential preserve in Brelsford-Gruber-Carlberg. Industrial use was scaled back to include only the Rinker Concrete Plant along the rail line. The coastline east of Flagler Boulevard was reserved for open space, with the exception of existing development just north of the Loftin Street bridge going to Palm Beach. Residential development in Northwest and the Garden District was emphasized, and the area west of the Seaboard Air Line tracks known as Clear Lake was also reserved for residential use. Okeechobee Boulevard east of Australian Avenue was proposed to be redesigned as a parkway entrance to the CBD. This recommendation, first made in the 1965 Milo Smith's plan, would direct as much traffic as possible from I-95 leading to downtown through this "Gateway." The CBD itself would grow by increasing densities and new building heights rather than expanding into the adjacent residential neighborhoods. Government Hill was viewed as an area that could develop internally with new county, federal, and state buildings and perhaps a convention center and performing arts center located on the site of Twin Lakes High.

Above all, the 1978 City Center Plan stressed preservation of the residential neighborhoods adjacent to the CBD. Stated in the final plan report:

> Where possible, existing concentration of single and two family dwellings should be protected. Multiple family development emphasizing low-rise characteristics should be encouraged between Tamarind and the FEC Railroad north of 1st street. The area bounded by 1st street, FEC Railroad, Palm Beach Lakes Boulevard, and SLC Railroad should continue to receive Federal Community Development Assistance to restore housing values, to encourage residential rehabilitation and to stimulate reinvestment or new investment by the private sector.[6]

The planners realized that even though the 1970 U.S. Census showed a 2,000 person decline from that in 1960, there remained some estimated 8,500 residents in 3,600 dwelling units in the City Center sub area. Retention and enhancement of this population was crucial to creation of an 18- to 24-hour pedestrian and activity life for downtown, as well as a market for everyday retail trade needs. Development of office and even residential uses on a conservative scale in the CBD core was a well-founded notion: with the exception of a few 10- to 14-story structures, most CBD blocks contained low-profile older buildings that could be redeveloped in office and residential high-rise structures. Competing office developments in the planning and development stage along the I-95 corridor from West Palm Beach south to Broward County further constricted the office market for downtown West Palm Beach.

West Palm Beach's planners had enough vision to see that a sketch plan for City Center was nothing more than a necessary first step. They wanted to take the next step — preparation of a detailed urban design plan showing buildings, open spaces, and streets. Stated the report:

> The City should pursue, with as much local talent as possible, the preparation of a more detailed comprehensive plan of the entire City Center Planning Subarea. The emphasis of this more detailed planning effort should be on the preparation of a design plan for the redevelopment and renovation of the Central Business District (CBD), and for development of the evolving governmental services center. The CBD portion of the study should produce a plan to identify how blocks and properties could be consolidated and redeveloped; how the CBD will relate to the waterfront; how Route 1 will be handled along the FEC

> Railroad; and how the entire plan will be implemented and monitored. The governmental services portion of the study should address how land in this area can be obtained to ensure that space will be available for further expansion of State and Federal governmental facilities. Important in this plan is not only the urban design criteria, but the public and private management and financial requirements necessary to carry out the plan.[7]

The entire West Palm Beach Comprehensive Plan was adopted by the City Council in May 1978. The planning department prepared RFPs for an urban design plan for City Center in late 1978 and early 1979. However, the recession of 1979–82 lowered tax revenues, and that, coupled with budget cuts in the City's Community Development Block Grant Program in the early 1980s due to Reaganomics, made it impossible for the city to fund such an effort. In 1982, the local elections produced, for the first time in the city's history, a majority of pro-development commissioners. The planning department, now working with depleted resources, moved ahead, and with the use of 1980 U.S. Census Data and survey information, produced in 1985 the new City Center Plan.

The City Center Plan of 1985: Bigger Is Better?

The 1980 Census showed West Palm Beach's population increasing from 57,375 to 63,108 without the benefit of significant annexation. Population in City Center declined from 8,147 in 1970 to 7,326 in 1980. With Palm Beach County growing by 125,000 people during this same time period, planners now saw an opportunity to extend the CBD core into the weakening residential neighborhoods to provide for what they thought were the space needs of a larger governmental and private sector workforce that would be the obvious result of this new growth. Even though employment in City Center dropped from 11,856 in 1965 to only 9,196 in 1984 while countywide employment during the same period rose from 90,000 to 275,000, West Palm Beach planners forecasted that at least an increment of future job growth could be directed back to the City Center.

These "forecasts" were not the efforts of carefully calculated market research efforts by city staff or economic development consultants, but educated "guesses" put forth initially by property owners in the CBD and their agents and endorsed by the pro-growth oriented City Commission and their planning staff. For example, the City Center Plan stated the following in the background data section:

> Several major office developments are under construction in 1985 in the City Center. An estimated 6,246 employees will be added to the 1984 total employment upon completion and lease out of these buildings. By 1986, approximately 13,200 employees will be working in the City Center.[8]

Even though these calculations allowed for a percentage of CBD employees leaving older quarters for newer ones and for a vacancy factor, we are not told what lease out actually means. Does it mean 100% occupancy? 90%? 80%? What vacancy rates would be reasonably expected by 1986? The net increase to 13,200 employees represents a guesstimate of the highest and most improbable order. Similar types of guesses without the presentation of supporting data appeared elsewhere in the final plan document.

During the early 1980s, despite the nationwide recession, a boom town spirit and mentality permeated southeast Florida, including West Palm Beach. Based on rapidly increasing population countywide and regionally, (the Dade, Broward, Palm Beach county area) planners and elected officials in West Palm Beach anticipated that an incremental portion of this growth could be directed to the City Center area without conducting detailed and carefully constructed studies to prove their assumptions. Some studies were questionable even from initial observations. For example, Table 11 of the City Center study showed that property values in the "downtown area," defined as those blocks within the new Downtown Development Authority's boundaries, jumped from 6.88% of total city valuation in 1983 to 10.19% in 1984, but noted only in passing that the Downtown Development Authority District's boundary expanded by over 50% between 1983 and 1984.

The 1985 City Center Plan's basic assumptions were (1) West Palm Beach and the City Center area would continue to experience growth, and (2) the City Center will become the city's primary location for new office and commercial development. A key projection developed by the staff and Real Estate Research Corporation anticipated a growth of commercial space in City Center from 1.2 million square feet between 1985 and 1989 to 1.9 million between 1990 and 1994.[9] As contrasted with the 1978 plan, City Center 1985's land use plan showed a "CBD" designation moving west across the FEC tracks halfway toward Tamarind Avenue between 2nd Street and Evernia. Actually, the 1985 City Center Plan was no plan at all. The CBD designation is shown on page 38 of the plan as one where a "diverse mix of uses will be allowed." It states specifically that *"the uses may include office, general commercial, government, residential, parks, and open space planned unit developments, mixed use developments and light industrial."*[10]

While the plan was specific in terms of tying the recommendations to needed capital improvements, just as rule R9j5 of the 1985 Local Government Comprehensive Planning and Land Development Regulation Act would call for the following year, it came up far short of the expectation posed by the 1978 study. There was no urban design plan. Uses proposed for inclusion within the CBD designation — office, general, commercial, open space, industrial, etc. — were not made site-specific. The residential neighborhoods — Brelsford-Gruber-Carlberg, Northwest, and the Garden District — were compromised, as existing residential blocks were to be replaced with either industry, the ubiquitous "CBD" designation, strip commercial, or a so-called new category of "office/residential." Still, despite the obvious contradiction between the desires of the 1978 plan and the 1985 version, West Palm Beach's Local Planning Agency and the City Commission passed the City Center Plan and incorporated it into the Comprehensive Plan as an "optional element" as per the 1975 and 1985 state Acts in December 1985. While the City Center plan met the 1985 Act's requirements — even those of R9j5 — it failed to anticipate another event that would greatly alter the City Center's shape. That event was the 1987 announcement of the "Downtown–Uptown" mixed use development project.

The Rise of Downtown–Uptown

As the recession of the early 1980s eased and money became more available at favorable interest rates, construction in the City Center took an upturn. Between 1984 and 1987 over 6,000 employees were added to the 1984 total working in City Center due to the construction of several new office buildings constructed in the old Central Business District area east of the FEC railroad tracks and in the Clear Lake District (see Figure 10), which was proposed in their 1978 plan for residential development.[11]

Residential construction took the form of high-rise towers on Lake Worth, despite complaints of staff planners who reminded elected officials that the 1978 City Center plan called for open space use on the waterfront.[12] Between 1980 and 1987, four towers totaling 600 units were built, the most prominent being the Plaza, a 224-unit luxury condominium. The Plaza was purchased by Donald Trump in 1987 as another of this tycoon's much heralded acquisitions, bringing the city reams of positive publicity.

In 1985, as contrasted to Trump's bombast, two investors quietly entered town and began to purchase City Center properties parcel by parcel using third parties to silently assemble entire blocks of land. By the end of 1986, these investors — Henry J. Rolfs, who had helped develop Tysons Corners in the Washington, DC area during the early 1970s, and David C. Paladino II, a local youngster who attended Rutgers University and was an accomplished jazz musician — had completed their work. With Rolfs now in his late 70s providing the technical expertise and Paladino having knowledge of the local market and family contacts for investment capital, these two finally assembled 21 entire blocks and 85 acres of land, over half of which was in the City Center area's Garden District. On January 22, 1987, Rolfs and Paladino announced the "Downtown–Uptown" project at a press conference. With site plans and a scale model showing buildings, streets, and open spaces, their development program, finalized in October of that same year, called for 3.7 million square feet of office space; 190,000 square feet of retail; 800 hotel rooms; and 425 residential units. By 1987 there was a total of 7 million square feet of office space in City Center, and the city had proposed a building cap of 10.4 million square feet. Therefore, this project would take up all of that space and then some (see Figure 12).[13]

Rolfs' and Paladino's announcement took West Palm Beach's governmental, civic, and business leaders by complete surprise. Downtown–Uptown would effectively shift the Central Business District from its present location to the southwest quadrant of City Center. It would wipe out the Garden District residential neighborhood, where just over 2,000 people lived (1980 U.S. Census), replacing it with a mix of office, retail, and hotel uses as Downtown–Uptown's residential component was planned for the portion south of Okeechobee Boulevard and outside of the defined City Center area.

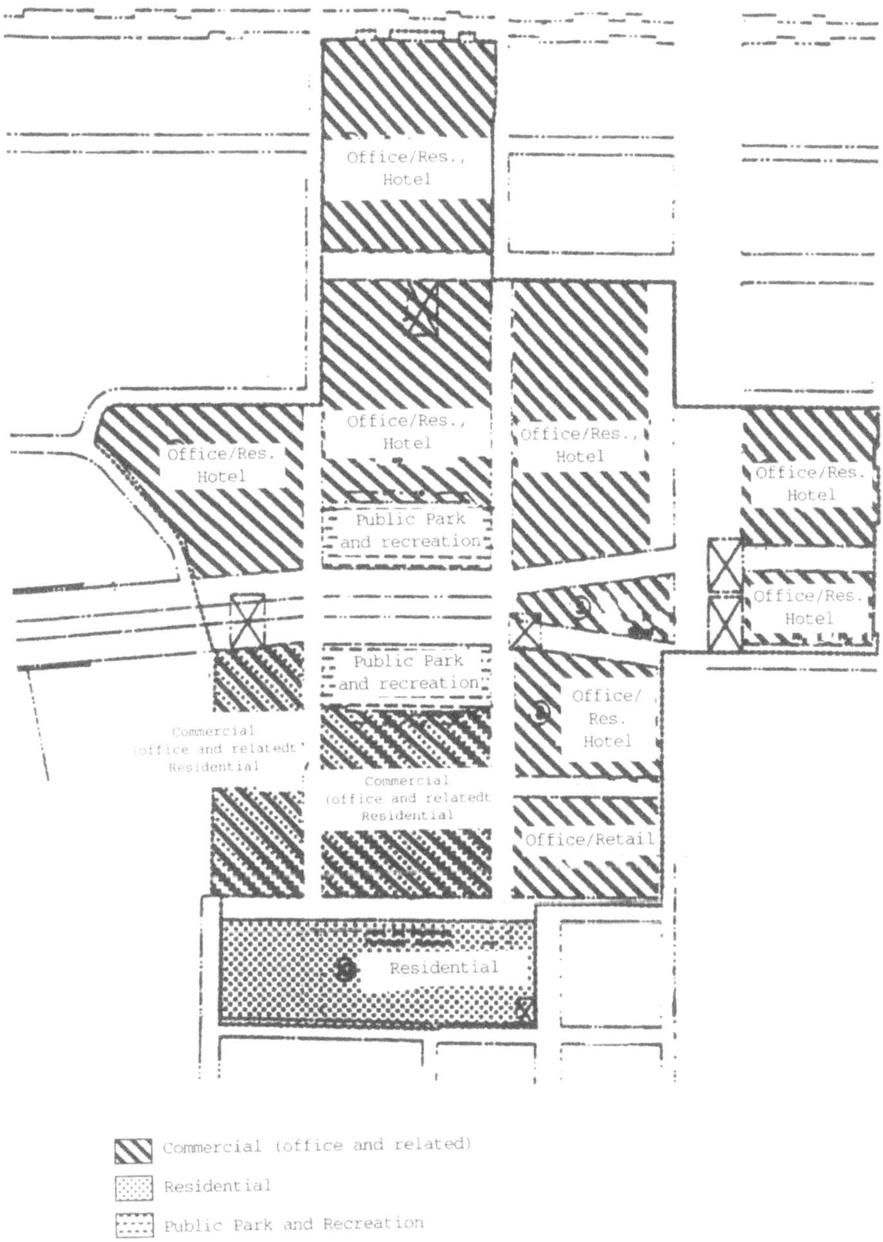

Figure 12. Proposed land use: The Downtown–Uptown project.

Stunned city leaders had only garbled responses to this development proposal. In a *Palm Beach Post News* article, City Manager Paul Steinbrenner was quoted as saying, "I've seen drawings of a park with buildings around it" ... "it runs north and south of Okeechobee" ... others had "no comment."[14] Before the collective city leaders could catch their breath, Rolfs and Paladino began to "constructively" evict tenants in the 450 dwellings remaining on their properties by turning off utilities. When tenants balked, the developers simply paid them to leave, with disbursements averaging $1,100 per family unit and $5,000 per business, 75 of which were on the site.[15]

A project of such massive scale would naturally generate a Development of Regional Impact study (DRI) as mandated by the 1972 Land and Water Management Acts referred to earlier in Chapter 3. That review would be conducted by the Treasure Coast Regional Planning Council (TCRPC) based in nearby Palm City. But Rolfs and Paladino were ready for this, too. Without waiting for the regular process to begin, they initiated their own DRI in September 1987.[16] As a result of the 1985 City Center Plan, West Palm Beach had begun work on a "Downtown Development of Regional Impact" study, but by late 1987 had only collected data and hired consultants. Now, with Rolfs and Paladino taking the lead role, the city government found themselves in the rather dubious position of having to follow these developers rather than lead the redevelopment process.

Placed in the embarrassing position of having to react quickly to Rolfs and Paladino, city government struck back with two documents released in January and February 1988.[17] The first, entitled *Present Issues: Future Visions,* prepared by the City Manager's Office, was essentially a "wish list" of 62 projects within the city proposed for development either by the public or private sector for completion during the 1990s. Eight of these projects were proposed for the City Center area, including a convention center, festival marketplace, public parking, a downtown bus loop system, renovations to the City Hall and City Library, in-fill housing programs, and of course, Downtown–Uptown. While the first seven projects were fairly well thought out in terms of total cost, city cost, local funding source, and state-federal private funding sources, the figures for Downtown–Uptown read, respectively, "unknown, $0, N/A, and private." After stating "no direct costs to the city are anticipated from this project" on page 48, the report went on to state, on page 49:

> It is recommended that City staff work closely with the TCRPC and other applicable government agencies to carefully analyze the potential impact of the proposed project on the City and the surrounding area, especially with regard to traffic circulation. In addition, the City should work with the developers of this project to retain at least the same amount of housing units as currently exists in the project area and to provide a full-service grocery store, as part of the project, if it is approved by the City Commission.

The planners had noted that: (1) there were 750 dwelling units in the project area in 1960 (U.S. Census) and although the number had declined to only 450 by the time of Rolfs' and Paladino's announcement in January 1987, given the policies of their 1978 and 1985 City Center Plans to manage residential use, 750 dwellings seemed to be a reasonable expectation; and (2) market studies incorporated into the 1985 City Center Plan forecasted a need for 1400–1600 new dwellings in that area between 1985 and 1994.

The February 1988 Evaluation and Appraisal Report for the West Palm Beach Comprehensive Plan prepared by the city's planning staff also addressed the Downtown–Uptown project, albeit in a somewhat oblique fashion. Quoted from page 4.3.4 of that document:

> Since adoption of the Plan in December 1985, the City has initiated discussion to develop a downtown convention center and festival marketplace. Five proposals have been received by the City to develop a Convention Center within the City Center area. In addition, the firm of Aaron Cohen Associates has been selected to determine if there is another site within the City suitable for the West Palm Beach Public Library. The existing site has been chosen as a potential site for a proposed festival marketplace. Several major roadway and parking facilities have been planned or are under construction and a 900,000 square foot expansion of the Palm Beach County Judicial Complex has been approved. A private sector development team has also announced plans for "Downtown–Uptown," an 80-acre development located on either side of Okeechobee Boulevard.

Therefore the city government's stated response to Rolfs and Paladino was essentially "go ahead and do it, but please, pretty please, put some more housing in." Downtown–Uptown was in violation of both the 1978 and 1985 City Center Plans. The 1978 plan called for residential use on every one of the 12 blocks within Downtown–Uptown's boundaries that lay within the City Center area, and the 1985 plan had designated 10 of the 12 blocks for residential or office-residential use. The Downtown–Uptown plan (see Figure 12) called for no housing on City Center's portion of its site. In 1987, despite the new construction downtown, there remained a 20% vacancy rate in office buildings constructed since 1970."[18] Given this substantial vacancy rate, *where were the studies that justified increasing the amount of office space in West Palm Beach outside of the old CBD area?* The city of West Palm Beach's planning department never commented on this either in official reports during the period of 1987–1989, and the studies justifying Downtown–Uptown never appeared.

During 1988 and 1989 Rolfs and Paladino, unhampered by serious questions about the feasibility of their development from the city's planning department, continued to relocate residents and businesses from their properties. By January 1989, the site was completely cleared.[19] In March 1989, Rolfs and Paladino entered into a joint venture with Tishman Speyer, the giant New York City–based real estate firm. The agreement called for Tishman to market the property to individual entrepreneurs.[20] The Development of Regional Impact study was completed by the Treasure Coast Regional Planning Council staff in late 1988 with data taken from Rolfs and Paladino, the City of West Palm Beach, and the Regional Council's files. The proposal was reviewed as per the 1985 Omnibus Growth Management Act. It was reviewed for compliance with the State Comprehensive Plan, the Regional Plan and R9j5. The staff found Downtown–Uptown in compliance with all three instruments, but as I was told in an interview with the planner in charge of DRI processing, "they met with the letter of the law but not with the spirit."[21] In the staff report, the Treasure Coast Regional Planning Council noted that traffic circulation would be a problem and that more housing and less office space was preferable. On January 20, 1989 the Regional Council approved the staff report, recommending that 1,000 units of housing be approved and that one-quarter of the project's nonresidential component be approved for completion by 1994, with the balance subject to approval after 1994. On February 13, 1989 these findings were transmitted to the City Commission of West Palm Beach, which voted on May 8, 1989 to accept the Regional Council's recommendations.[22] Rolfs' and Paladino's response was that they did not foresee a market for 1,000 units of housing on their site but that they would be interested in contributing to a housing fund in lieu of building a "negotiated" number of units. Rolfs' and Paladino's position was attacked in editorials by the *Palm Beach Post,* which echoed sentiments of the planning department and city elected officials, who wanted to maximize housing as a reuse in Downtown–Uptown.[23] If the developers were to have their way, the project could then become 100% nonresidential.

The Northwest Neighborhood Responds

With the Garden District eliminated by Downtown–Uptown, City Center was left with only two residential neighborhoods — Breslford-Gruber-Carlberg and Northwest. Brelsford-Gruber-Carlsberg had a population of 925 residents in 1970 and 927 in 1980. Luxury high rises on Lake Worth had replaced some single-family units that were lost to encroachment by commercial uses from the CBD core on the south and medical-related office uses from Good Samaritan Hospital just north of this area. Local residents and city staff had generally agreed that the 1985 City Center Plan provided for their needs and that of expanding nonresidential activities.

However, Northwest was an entirely different matter. Once the center of residence for West Palm Beach's African-American middle class, most of these residents moved out during the 1970s and 1980s when federal open housing legislation provided for greater choice. However, regardless of place of residence, most blacks in West Palm Beach *cared* about the Northwest. Almost all of the black community's churches were located there — as was Palmview School, the importance of which was noted earlier in this chapter. Residents present and former realized the neighborhood's potential. It was right next to downtown but was protected from through vehicular traffic by the railroad. Located on the Coastal ridge, the Northwest neighborhood offered spectacular water views. Given this, coupled with direct proximity to Palm Beach Lakes Mall and I-95 only two miles away and the new Tri County Commuter Rail Terminal located on the neighborhood's periphery, Northwest would be perfect for upper middle class (white) residence. It could very well be a case once more of "Negro Removal," just like the days of urban renewal in the 1950s in places such as Baltimore, Washington, DC, Chicago — and Miami.

Community leaders were well aware of this possibility and watched carefully during the 1970s and 1980s for city plans that would accelerate such a process. There was some comfort in the fact that both the 1978 and 1985 City Center Plans called for maintaining the neighborhood by use of in-fill housing, housing rehabilitation, and expanded open space for Palmview School.[24]

However, in March 1987 a developer proposed that a county records storage building be erected on a 6.5-acre triangular site just opposite the FEC railroad tracks along Rosemary Avenue (see Figure 11). The site, once a thriving commercial area, had been gradually cleared by the city's Building Department after the obsolete stores slowly closed and their structures had deteriorated. Neighborhood activists had viewed the site as a resource for about 100 units of affordable housing. Not only were they displeased by foreclosure of such a possibility, but the proposed use was to be an 8-story windowless structure completely blocking residents' view of the downtown skyscrapers. Local residents dubbed the project "The Wall," and fought a pitched battle against it in public hearings on the zone change application, and the developer dropped his plans for the records center.[25]

The residents were led by U.B. Kinsey. Kinsey moved to West Palm Beach with his family in 1932 from rural north Florida. Graduating from Industrial School in 1938 — the area's only high school for blacks — Kinsey attended Florida A&M University, graduating with honors in 1942. After service in World War II, he joined the Palm Beach County Schools as a teacher in 1946. Principal of Palmview Elementary school since 1960, Kinsey was a strong advocate for the school and the Northwest neighborhood. Known as a perfectionist, Kinsey would walk the halls of the school making sure that children were in class and quietly learning. He would inspect the meticulously landscaped grounds making certain that maintenance men were doing their job as well. He commanded respect both in the neighborhood and at City Hall. When Mr. Kinsey spoke, everyone listened, despite his slight stature and calm quiet voice.

By 1987, the Northwest neighborhood had strong allies on the City Commission and the city Planning Department's staff. Two of the five city commissioners were black — James O. Poole, a retired school administrator, and Sam Thomas, a native of West Palm Beach and a prominent attorney. These two were joined by Pat Pepper Schwab to form a 3–2 liberal majority, one sympathetic to minority group issues. Two African-American women had recently joined the Planning Department's professional staff. Shirley Simpson-Wray was the city's community development coordinator. Originally from Boston, Ms. Simpson-Wray held a master's degree in planning from MIT and was a full member of the American Institute of Certified Planners. Sharon Jackson, a graduate of the Georgia Institute of Technology's master's program in City Planning and a former student of mine when I taught there in 1976–77, had almost 10 years of experience as a professional planner when she joined the West Palm Beach planning department.

The Downtown Development of Regional Impact study, initiated by the planning department as part of the 1985 City Center Plan, called for neighborhood plans for the three residential areas within the central core. With the help of Simpson-Wray and Jackson, the monies for a new neighborhood plan for Northwest were made available in the form of a consulting contract to Thaddeus Cohen, a black architect/planner with offices in Delray Beach. Cohen asked me to join his team as an adviser. I found in my inventory studies that, although the neighborhood's population had dropped from 4,000 in 1970 to 3,100 in 1980, population had stabilized by 1988 as demolition permits had dropped to almost zero, and the student population of Palmview School was higher than in 1980. Several local contractors had expressed interest in developing small multi-unit complexes if federal subsidies could be made available. This, coupled with the neighborhood's advantages, amenities, and several strong indigenous community organizations, meant that a revitalization strategy was called for. I noticed that all of the major historic buildings lined up on one street — Division. Running south from 11th Street and Palmview School were the Payne Chapel AME Church, constructed in 1928 in Romanesque Revival style; Tabernbade Baptist Church, built in a modified gothic style in 1926; the Augustus/Hazel house, a mansion built in 1915 as the office and residence of West Palm Beach's first black architect; and Friendship Baptist Church, built in 1930 and styled in the Spanish Colonial/Pueblo vernacular. Using historic preservation as a catalyst for neighborhood upgrading, Cohen and I proposed that Division be designated a "historic" avenue with new brick paving, 19th Century style street lights, and related landscaping. Other proposals included expansion of Palmview School's site for additional outdoor open space, use of the "Wall" site for family housing, and in-fill housing construction on vacant lots. The plan, enthusiastically supported by neighborhood residents, was accepted by the Planning Department in June 1989 and incorporated into the "new" City Center Plan of 1989. By the end of 1991, the streetscape project for Division Street had been funded for $300,000 by the city government, a mixed use housing and commercial development was underway on the old "Wall" site, and the Northwest neighborhood seemed on its way back.

The Gardenia Walk HODAG

In 1984, the City of West Palm Beach was awarded a Housing and Development Action Grant (HODAG) in the amount of $2 million by the U.S. Department of Housing and Urban Development for construction of the 100-unit Gardenia Walk Apartments and renovation of the 30-unit Hibiscus Garden Apartments. Both developments were in the Garden District, and when completed this project would have provided a total of 230 Section Eight assisted housing units for low- and moderate-income families and the elderly. The City of West Palm Beach accommodated these projects by designating the area bounded by Fern, Florida, Hibiscus, and Georgia as a Downtown Planned Unit Development (or DPUD) in 1985.

The Gardenia Walk project was formulated by Gardenia Walk Associates, a housing development corporation that consisted of several residents of the Garden District area and individuals who were associated with Neighborhood Housing Services, which had been active in the area since the late 1970s. Gardenia Walk Associates felt that if their new construction project was successful, it could serve as a catalyst for revitalization of the entire Garden District. The Hibiscus Arms apartments, built in 1926, had been placed on the National Register of Historic Places. Restoration of this building would have provided publicity that would in turn focus public attention on the Garden District and its possibility for revitalization.

Gardenia Walk Associates appealed to the City of West Palm Beach for assistance in land acquisition and provision of infrastructure in the form of upgraded water and sewer lines, street paving, and landscaping. Estimated total cost was $12–13 million, and the HODAG grant would only cover the down payment required for mortgage monies. The city refused to help even by declaring the DPUD area an eminent domain district where the site could be acquired based on appraisal value rather than the exorbitant speculative prices asked by the property owners.

Because of the city's reluctance to help out, Gardenia Walk Associates lost the HODAG grant, and those monies were transferred to another new construction project known as Cypress Run, located on the city's northwest edge. Meanwhile, the Gardenia Walk Apartments' site was acquired by Rolfs and Paladino in 1986 for an undisclosed sum. Rolfs and Paladino also acquired the Hibiscus Garden Apartments and, despite the fact that this structure was on the National Register, demolished it.[26]

The 1989 Center City Plan

During most of 1987 and 1988, the West Palm Beach Department of Community Development and Planning was busy preparing the city's new comprehensive plan. This document, consisting of 16 chapters and a total of over 800 pages, contained all of the mandated elements of the 1985 state act, including future land use, traffic circulation, mass transit, port, aviation, and related facilities, housing, sanitary sewer, solid waste, drainage, potable water, and natural groundwater, aquifer recharge, coastal management, conservation, recreation and open space intergovernmental coordination, and capital improvements. An optional element, the new City Center Plan, was strikingly similar to its 1985 counterpart in terms of overall presentation format, goals and objectives, and implementation strategy. However, there was one major difference, even though it was subtly disguised.[27] It was Downtown–Uptown.

The only reference to Downtown–Uptown in the 1989 City Center Plan was a partial paragraph on page 16-6. Quoting from page 16-6 of that plan:

> The acquisition by one developer of most of the district (the Garden district neighborhood), plus land south of Okeechobee Boulevard presents the opportunity for redevelopment of the entire area as a commercial and residential gateway into the City. Additional lands south of Okeechobee Boulevard...is now incorporated into the City Center Element and with the old Garden District form the new Garden/Gateway District.

When the West Palm Beach Comprehensive Plan was finally passed by the City Commission on November 20, 1989, Downtown–Uptown was shown on the official plan map as "CBD," that same noxious designation used in the 1985 City Center Plan, which could include *"office, general commercial, government residential parks and open space Planned Unit Developments, Mixed Use Developments, and light industrial"* (pp. 16–35, 1989 City Center Plan). As shown by Figure 10, the 1989 plan simply extended the "CBD" use west and south from the original downtown. The Garden District, which housed 1,189 residents in 1970 and 1,062 in 1980 according to the U.S. Census, had been cleared by Rolfs and Paladino, and the 1989 plan permitted this entire neighborhood to be rebuilt as 100% nonresidential. The major question raised publicly in city hearings on the plan was: "Is a plan using up all of the proposed office space to be built over the next 20 years in downtown West Palm Beach feasible?" The only explanation was on pages 16-31 of the plan report, stating that, based on projection by the city staff and Real Estate Research Corporation:

> Employment projections were calculated by applying an average ratio of space per employee to the office and commercial space projections. By 1994, 13,300 new employees will be added to the 11,450 existing and committed employees for a total of 24,750. By 2010, the total employment will be approximately 37,300.

This projection has to be questioned in the light of several pertinent facts. First of all, while 4 million people resided in the southeast Florida megalopolis by 1990 (Dade, Broward, and Palm Beach counties) and this super area's centers are Miami, Fort Lauderdale, and West Palm Beach, the latter city is at the northern end of the metropolis and cannot draw from both north and south, as is the case for Fort Lauderdale. More important, growth is toward the west, past the Florida Turnpike, and in recent years many "edge cities" have been constructed west of I-95 in Southeast Florida to meet the demand for office and related activities. These complexes include not only the Village of Palm Beach Lakes, but the IBM center in Boca Raton, Cypress Creek in north Broward County, and Miami Lakes–Hialeah. Between 1985 and 1989 several development proposals for office centers near downtown West Palm Beach were put forth. The two largest were Centerpark, adjacent to the Palm Beach International Airport, calling for 1.4 million square feet of office space, 75,000 square feet of commercial, and 225,000 square feet of hotel; and Northpoint, at the Intersection of I-95 and 45th Street with a total of 1.4 million square feet of office, commercial, industrial, and hotel use. Two other major proposed developments in the I-95–45th Street area include the 45th Street business park, with 1.3 million square feet of industrial and office use; and Metrocentre Corporate Park, with 570,500 square feet of office, hotel, and retail use. All four of these developments received at least partial DRI approval by 1989, and their total of 5 million square footage is located directly adjacent to I-95, while Downtown–Uptown lies one mile away from the nearest I-95 interchange.[28] The only locational advantage for Downtown–Uptown would be for firms doing business with city, county, state, and federal agencies, which is a fraction of the commercial office market. *Given all of these competing centers with better locations, how does one justify Downtown–Uptown?*

Although Downtown–Uptown's site was cleared by January 1989, and the City Center Plan released in November of that same year, the description of the Garden District in terms of overview, zoning analysis, and neighborhood trends *is virtually the same as the 1985 plan.* The 1985 goal for the Garden District was *"the city shall retain and enhance the Garden District neighborhood and encourage a variety of housing opportunities for all income levels"* (pg. 120). The 1989 goal for this neighborhood read on page 16-108 *"the city shall enhance the expanded Garden/Gate District neighborhood by encouraging a variety of housing opportunities along with the commercial redevelopment of the district."* The only problem, though, was that the city forgot to mention that by 1988 the Garden District no longer existed, and there was no firm commitment on *anyone's* part — city government or the Downtown–Uptown developers — to provide 450 units of new housing, or any units for that matter, to replace those that were demolished. Both the Existing Land Use and Future Land Use maps were identical in the 1985 and 1989 plan reports, which was ironic because by 1989 existing land use in the Garden District was 100% vacant and the proposed land use was 100% nonresidential — quite different from the 1985 scenario. The fate of the Gardenia Walk and Hibiscus Garden Apartment project was not mentioned except for a cryptic note on page 16-106, stating, "Hope for new residential projects have not materialized." The final irony was that Objective 1, just under the goal statement, read *"Increase the city support for neighborhood improvement organizations." What* neighborhood organizations? Thanks to Downtown–Uptown, there was no neighborhood!

Despite these discrepancies, the entire Comprehensive Plan, including the City Center element, sailed through the Florida State Department of Community Affairs (DCA) review process. DCA ruled that the Comprehensive Plan was consistent with the State Plan, the Treasure Coast Regional Plan, and R9j5 stipulations. As for the City Center Plan, in the words of the head of the Treasure Coast Regional Planning Council's DRI section, while it met with the letter of the law, it certainly didn't meet with its spirit. While acceptable to state agencies, in this author's opinion, it is certainly questionable as to whether it was acceptable within the realm of good planning practice as embodied in the American Institute of Certified Planners' Code of Ethics.[29]

The Fall of Downtown–Uptown

By 1989, Rolfs and Paladino were riding high, wide, and handsome. Downtown–Uptown was rolling along with virtually 100% clearance of the site and relocation of tenants. The management agreement with Tishman and Associates of New York City, if implemented, could generate millions for the pair and their associates once vacant land was sold off.

But by late 1989, the bubble had burst. Due to the impending recession and downturn in the commercial construction market, Tishman pulled out of the agreement on October 27, 1989.[30] At least that was the newspaper account; it could be that Tishman, after additional investigation, decided that, given the competing projects along I-95, which were further along in the development process, Downtown–Uptown wasn't feasible even on a long-term 10–15 year basis. After Tishman's withdrawal, Rolfs' and Paladino's empire began to crumble like a house of cards. Short-term investors who held mortgages on Rolfs' and Paladino's properties began to file foreclosure suits.[31] By 1992, this project, for all intents and purposes, was dead. However, the land they bought and cleared remained scarred and desolate like some intergalactic moonscape.

Some good did come from Downtown–Uptown. In 1991, the Kravitz Center for the Performing Arts was completed on a one-block site in the old project's southwest corner. On December 12, 1991, the City of West Palm Beach approved plans for a festival marketplace on a portion of Downtown–Uptown, and a $2.5 million purchase of land was made to widen Okeechobee Boulevard, which was essentially a partial bailout to the Downtown–Uptown developers. The entire Downtown–Uptown site could become a mini Central Park, with not only the Performing Arts Center and festival marketplace but a small-scale Convention Center as well. Perhaps the 450 units of housing lost to demolition for Downtown–Uptown can be replaced by at least 1,000 units in all price ranges, as recommended by the Treasure Coast Regional Planning Council in their Development of Regional Impact Review.

But if Downtown–Uptown does eventually come to some net good, the West Palm Beach Department of Community Development and Planning will have a hard time taking credit for it. City government rolled over and approved Downtown–Uptown by looking the other way while initially introducing a 1985 City Center Plan that was no plan at all, with the notorious "CBD" designation, then compounded the non-planning with a 1989 City Center Plan that all but ignored the reality of Downtown–Uptown and the problems posed by this development. Actually, the 1978 Comprehensive Plan's City Center element was more realistic, and certainly more humane.

Planning: Its Definition; What Should Have Been? Why the Succession of Plans?; Intervention of Mandated Planning

Planning Defined

There are perhaps as many definitions of planning as there are planners. Webster's Encyclopedic Unabridged Dictionary of the English Language defines "planning" as "a devised scheme; a way of executing an act; a method of arrangement; an aim or project." Scholars of urban and regional planning have over the years developed their own definitions of planning as applied to public/private decisions in the urban and regional environment. Andreas Faludi defined planning as "the application of scientific method — however crude to policy making."[32] Paul Davidoff defined it as "a process for determining appropriate further action through a sequence of choices."[33] Edward C. Banfield defined planning as "the process by which he selects a course of action (a set of means) for the attainment of his ends."[34] John Freidmann, who, in my opinion, is the nation's top scholar in planning theory by the early 1990s, defined planning as "a way of managing the nonroutine affairs of the city...including... economic expansion, social welfare, education, housing, public transportation, public health, culture and recreation, land use control, and urban design."[35]

I especially like Freidmann's definition, because it is comprehensive, precise, and directly to the point. Central to this definition is the term "managing." If we carry this notion to its logical conclusion, especially including all of the areas mentioned by Friedmann, we must conclude that the ultimate responsibility for "managing the nonroutine (routine would be day-to-day management, such as personnel and financial administration) affairs of the city" must lie with *elected officials.*

Elected officials can either lead by example or follow the whim of public opinion. Leading by example is best shown by Florida governor Reubin Askew's speech in 1971 to a conference of state political business and institutional leaders, when he said, "It's time we stopped viewing our environment through prisms of profit, politics, and geography or local and personal pride."[36] It is clear that in the case of the City Center Plans of 1985 and especially 1989, the elected officials in West Palm Beach were doing less than "managing the nonroutine affairs of the city." No one put a gun to Rolfs' and Paladino's heads and made them quietly assembly 21 blocks and 80 plus acres of prime urban land at a cost of millions (mostly borrowed). The 1985 City Center Plan called for primarily residential use in the Garden District. Zoning was mostly R-5 — multi-family. Elected officials and the planners who serve them could have politely told these developers, "Your project is in violation of our City Center Plan of 1985, which was itself derived from the 1978 Comprehensive Plan prepared in accordance with state law. Please redesign your project and come back to us."

In the case of City Center, it was extremely important for elected officials to be advocates for citywide interests. First of all, they should have recognized Downtown–Uptown for what it was — *a carefully developed land speculation scheme.* Similar projects were on the front burner in other parts of the city — namely, at interchanges along I-95. The developers presented no market studies that justified such a project, especially given (1) the existing vacancy factor in downtown West Palm Beach, and (2) the competing projects along I-95 in West Palm Beach and regionally, additional projects in Broward County (Cypress Creek) and Miami Lakes–

Hialeah. It was very important for elected officials to be advocates for downtown because the citizens of West Palm Beach cared about their neighborhoods and trusted others to take care of the CBD. When sub area plans were prepared, residents participated with vigor and commitment. Even in the City Center neighborhoods of Brelsford-Gruber-Carlberg and Northwest, citizens were strong advocates for their communities. The Garden District's residents by 1985 were elderly or recent Haitian immigrants. They were not organized, but they were still human beings who needed advocates.

When the developers entered into a partnership with Tishman Speyer, their intentions were clear to all who cared to look without blinders. They had bought the land and cleared it. Now it was time to sell to another party (or parties) who would then take the long-term risk. It was important for West Palm Beach elected officials and long-time "establishment" leaders to see past the slick illustration and colored slides and get to the hidden agendas. *Why didn't they?*

What Should Have Been

One reason elected officials couldn't just say no was that the 1985 City Center Plan was seriously flawed. The 1978 City Center Plan called for an urban design three-dimensional study to "identify how blocks and properties could be consolidated...how the CBD will relate to the waterfront...and how the entire plan will be implemented and monitored." Rather than follow through with this well-intended and well-developed thought, the 1985 plan took a step backward. A two-dimensional map was the only offering, and its "CBD" designation, permitting everything from residential to light industrial was much less clear than the proposed land use sketch offered by the 1978 City Center Plan.

The design study called for in 1978 was not the reinvention of the wheel but a practice tried-and-true in downtown revitalization. Just two of the immensely successful CBD projects, namely Philadelphia and Baltimore, came from this type of study. Philadelphia's planning director, Edmund Bacon, was not only a practicing planner and administrator during the period of 1950–1970, but a superb scholar as well. His book *Design of Cities* written in 1967[37] was a summary of the plan for Center City, Philadelphia's downtown, and the original townsite for the city designed by William Penn in 1684. Philadelphia's Center City Plan, prepared by Bacon and his staff and completed in 1958, was an urban design scheme showing the precise location of buildings and open spaces in relationship to streets and pedestrian ways. Bacon's plan resulted in the redevelopment of Center City in the period of 1960–1980. Today, Center City is an international showplace of a core developed as a result of "managing the nonroutine affairs of the city." Undoubtedly, the West Palm Beach staff who put together the 1978 City Center Plan were following Bacon's lead.

Baltimore has received virtually universal acclaim for downtown revitalization. James Rouse's magnificent Harborplace relates well to the office complex of Charles Center and both relate well to Camden Station, which was redeveloped as an industrial park office and residential complex, including a new stadium for the Baltimore Orioles. But Baltimore's CBD renewal was not by chance. In the early 1960s, under the leadership of then mayor Theodore McKelden, a group of local business and civic elites, tired of their city — thought of as a stepchild to Philadelphia and Washington, DC — decided to make their hometown a world-class city. This group, known as the Greater Baltimore Committee, hired a top urban design consultant to prepare a three-dimensional plan for the entire downtown core area (see Figure 13). When I came to Baltimore in 1965 to be project planner for the Gay Street I neighborhood — located just outside of the CBD, but within the downtown core — the urban design plan for that core had just been completed. Between 1965 and 1985 virtually everything proposed by this downtown plan had been built...according to the plan, not the whim of developers. For instance, Harborplace was actually part of the Inner Harbor area study calling for a combination of hotel, retail, and residential uses on what was in the early 1960s decaying wharves and warehouses. Once the plan for the Inner Harbor was complete, the Greater Baltimore Committee contacted Rouse and showed him just where his exciting Harborplace should sit. In West Palm Beach, Rolfs and Paladino announced their plans and then elected officials and the city's private leadership followed obediently like trained dogs on a leash!

Why the Cavalcade of Plans?

Of the three City Center Plans produced and adopted over an 11-year period, the first one, prepared in 1977–78, was the best. If the planners and elected officials had heeded that plan's advice and prepared a three-dimensional urban design scheme like those done in Baltimore and Philadelphia by 1985, potential developers and investors, including Rolfs and Paladino, *would have known exactly where to put their projects.* Then elected officials could have stated an authoritative NO! to the Downtown–Uptowns of this world when the projects didn't fit the plan.

One response to a question by citizens and news media as to why developers were calling the shots came from West Palm Beach vice mayor Pat Pepper Schwab. Ms. Schwab, who at the time was married to a prominent local architect-developer, was quoted as saying:

> We have leaders but they're not providing leadership. We react instead of being leaders which means being out there in front painting the vision.[38]

In West Palm Beach, the elected officials and their administrators are well-educated, enlightened, and generally progressive. The planning staff has a long-held reputation for competence, as evidenced by a long-standing positive record of comprehensive planning, including sub area mid-range planning tied to capital improvements programming. Up until 1981, West Palm Beach's elected officials were, for the most part, conservatives and traditionalists who did not support rapid growth, especially that initiated by outsiders. In 1981, a cadre of pro-development forces captured City Hall by electing a 3–2 majority to the City Commission. The pro-development forces have controlled city government ever since, but without a clearly defined leader. Perhaps the past lack of leadership is due to an old governmental structure, which is the mayor/council form. "The mayor" and "vice-mayor" were first and second among equals. The mayor and vice-mayor positions rotated among the five members of the city council, all of whom were elected at large. In the mayor/council form of government — used in Baltimore, Philadelphia, New York City, Chicago, Detroit, Los Angeles, San Francisco, etc. — the mayor is not a part of the City Council, and therefore the news media and citizens in general look to that person for leadership in terms of citywide issues, including downtown development. In Philadelphia, while Edmund Bacon was the planner, Harrison Dilworth was the mayor who pushed Bacon's Center City Plan. In Baltimore, Mayor Theodore McKelden helped to initiate the process; then his successors Thomas Del Assandro II and William Schaeffer provided the political leadership to fully implement the plan.

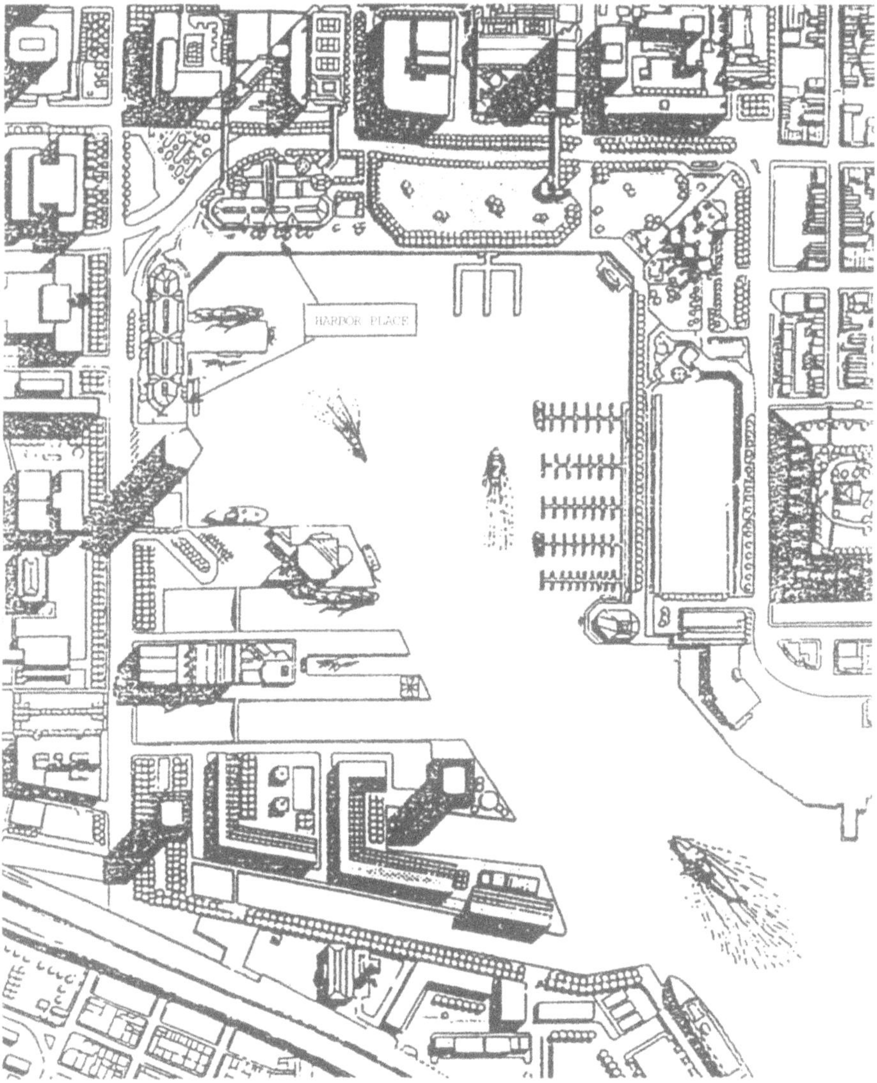

***Figure 13**. Example of an urban design plan: Baltimore's Inner Harbor, 1965. Source: Baltimore, Maryland Housing and Urban Renewal Administration.*

In 1991, West Palm Beach voters changed the form of government to that of strong mayor. In November 1991, Nancy Graham became the first mayor elected under that government.

The Role of State Mandated Planning

Both the 1975 and 1985 state Acts were designed to make units of local government plan in a rational comprehensive manner in order to successfully manage growth. Did they help to do this in West Palm Beach? This author doesn't think so. West Palm Beach prepared comprehensive plans even before 1969, when the State of Florida passed enabling legislation permitting local government to prepare and adopt plans. Milo Smith's 1965 plan included basically all of the elements required by the 1975 legislation, and its CBD plan and the citywide capital improvements element was an innovation for Florida. As mentioned earlier in this chapter, West Palm Beach's Planning Department had an active advanced planning process during the late 1960s and early 1970s, so when LGCPA came in 1975 only the format needed to be adjusted to prepare the 1978 Comprehensive Plan.

The 1989 Comprehensive Plan and the City Center Element were approved by DCA, as the plan was indeed consistent with state and regional plans and met the requirements of Rule R9j5. Downtown–Uptown was approved as a result of the Development of Regional Impact process. But was the City Center Element a *good* plan? It endorsed a development proposal, Downtown–Uptown, that violated the City Center Plan of 1985, without providing conclusive data to show that this 1985 plan's "predominantly residential" designation for the Garden District neighborhood should have been reversed. By no means can this action be termed "good" planning practice. Still, it met with the letter of the 1985 state law.

Conclusion

The 1975 Local Government Comprehensive Planning Act and its successor in 1985 may have lit a fire under those Florida communities that hadn't thought of planning as a way to harness growth. But given the cavalcade of plans for West Palm Beach's City Center area prepared between 1978 and 1989, it seems to have had little or no effect in terms of impacting the *big* decisions, namely, maintaining the integrity of neighborhoods such as the Garden District with its 1,000 former residents, and approving or disapproving major development projects, such as Downtown–Uptown. The 1985 legislation, with the minimum criteria rule of

R9j5, is fine for those communities that don't know how to plan; but for sophisticated cities such as West Palm Beach, with well-educated elected officials and competent planning staffs, it has minimal effect. No set of rules can change behavior. If leaders don't want to plan rationally, no legislation is going to force them to do so. To use the old adage, "You can lead a horse to water, but you can't make it drink."[39]

Notes

1. Edward Durr. *Some Kind of Paradise: A Chronicle of Man and the Land in Florida,* William Morrow and Company (New York), 1989.
2. Donald W. Curl. *Palm Beach County: An Illustrated History,* University Presses of Florida (Gainesville), 1986, pp. 37, 38.
3. City of West Palm Beach Comprehensive Plan (1989), p. 5-4.
4. Milo Smith and Associates. *West Palm Beach Comprehensive Plan* (1965).
5. Twin Lakes High School was formerly the all-white Palm Beach High School, opened in 1912 to serve most of the county's white population. When integration came to Palm Beach county in 1971, Palm Beach High was merged with all-black Roosevelt High School and renamed Twin Lakes. Even though Twin Lakes was added onto several times after 1912, by the early 1970s it had deteriorated and was designated for closing before 1990. Its site was seen by planners as an opportunity for a cultural center and a new countywide secondary school for the performing arts.
6. West Palm Beach Planning and Community Development. *City of West Palm Beach Comprehensive Plan* (May 1978), pp. 11, 41.
7. Ibid., pp. 11, 40.
8. West Palm Beach Department of Planning and Development. *The City Center Plan* (December 1985), p. 15.
9. Ibid., p. 25.
10. Ibid., p. 38.
11. West Palm Beach Department of Planning and Development. *The City Center Element of the West Palm Beach Comprehensive Plan* (November, 1989), p. 16-15.
12. Ibid., pp. 16-17.
13. *Palm Beach Post.* "Developer Seeks to Create City Within a City," p. 1A (January 23, 1987). *Palm Beach Post.* "Millionaires Unveil Plan for City Within a City," p. 1A (October 5, 1987).
14. Ibid.
15. Ava Van De Water "Downtown/Uptown Site Ready for Wrecking Ball," *Palm Beach Post,* p. 47, "Outlook" Section (January 29, 1989), *Palm Beach*

Post, "Buildings in Shadow of Wrecking Ball: Residents Make Way for Downtown/Uptown," p. 1A (March 31, 1989).

16. Larry Aydlette, "Downtown Plan Ambitious," *Palm Beach Post* (October 5, 1987), p. 1A.
17. West Palm Beach: Office of the City Manager, *Present Issues: Future Vision,* (January, 1988) and *An Evaluation and Appraisal of the City of West Palm Beach Comprehensive Plan* (February, 1988).
18. Interview with Shirley Wray, West Palm Beach Community Development Coordinator (February 17, 1989).
19. Op. cit. Note 15 (January 29, 1989).
20. Scott Saef, "Downtown/Uptown Gets Developer; Tishman Speyer Joins 75 Acre Project," *Palm Beach Post,* p. 1B, March 11, 1989.
21. Interview with Sally Black, Chief Planner, Development of Regional Impact Section, Treasure Coast Regional Planning Council (January 9, 1992).
22. *Palm Beach Post,* "Downtown/Uptown Changes Approved" (Tuesday, May 9, 1989), p. 2B.
23. *Palm Beach Post,* "Housing is Essential: Downtown/Uptown Residential Units Important," Opinion Section p. E2 (January 29, 1989). Also see *Palm Beach Post,* "Money No Substitute: Downtown/Uptown Cash Must Go Toward Housing," Opinion, p. 1F (March 27, 1989).
24. Op. cit. Note 18.
25. Interview with V.B. Kinsey, Northwest Neighborhood Civic Activist and principal of Palmview School (February 19, 1988).
26. Interview with Shirley Wray and Sharon Jackson, West Palm Beach Department of Planning and Community Development (January 10, 1992).
27. On almost all the 36 maps included in this 135-page statement was the wording at the bottom "Note: Figure to be revised to include the Downtown/Uptown project." Maps reflecting Downtown–Uptown were not printed in the text. The only map showing Downtown–Uptown's impact on land use was in the office. Source: Interview with Rick Green, West Palm Beach Planning Director (January 8, 1992).
28. West Palm Beach Comprehensive Plan (November 1989), pp. 5, 8.
29. The American Institute of Certified Planners Code of Ethics and Professional Conduct was first adopted in 1979 and amended through 1987. One clause that might have been violated by the West Palm Beach Planning Department is that of *The Planners Responsibility to the Public.* Section A3 reads, "A planner must strive to provide full, clear, and accurate information on planning issues to citizens and governmental decision-makers." The misrepresentation of existing land use and future land use for the Garden District in the 1989 City Center Element of the West Palm Beach Comprehensive Plan is, in this author's opinion, a possible violation of

professional ethics, despite the disclaimer "Note: Figure to be revised to include the entire Downtown/Uptown project."

30. Elisha Williams, "Downtown Co-Developer Backing Out," *Palm Beach Post* (October 28, 1989), p. 1A.
31. *Palm Beach Post,* "Bank Sues Downtown/Uptown Over Land," p. 22B (June 28, 1990); *Palm Beach Post,* "All Downtown/Uptown Loans in Default," p. 12 (November 1, 1990); *Palm Beach Post,* "Downtown Lender Sues," p. 12B (December 1, 1990); *Palm Beach Post,* "Downtown/Uptown Partners Sued Again," p. 11B (December 4, 1990); *Palm Beach Post,* "Downtown/Uptown Faces Two More Suits," p. 6 (December 24, 1990); *Palm Beach Post,* "Downtown/Uptown Faces More Suits," p. 7B (January 12, 1991); *Palm Beach Post,* "Lawsuit Filed Against Downtown/Uptown," p. 11B (February 20, 1991).
32. Andreas Faludi. *A Reader in Planning Theory,* Pergamon Press (New York), 1973, p. 1.
33. Paul Davidoff and Thomas A. Reiner. "A Choice Theory of Planning," *Journal of the American Institute of Planners,* Vol. 28 (May 1962).
34. Edward C. Banfield. "Ends and Means in Planning," *International Social Science Journal,* Vol. X1, No. 3 (1959).
35. John Freidmann. "A Response to Altschuler: Comprehensive Planning as a Process," *Journal of the American Institute of Planners,* Vol. 3 (August 1965).
36. Luther Carter. *The Florida Experience: Land and Water Policy in a Growth State,* Johns Hopkins University Press (Baltimore), 1975, p. 125.
37. Edmund N. Bacon. *Design of Cities,* Viking Press (New York), 1967.
38. Ava Van de Water. "On the Eve of a Dramatic New City, *Palm Beach Post,* p. 1 (November 15, 1987).
39. Some individuals would assume that the Downtown–Uptown developers did the city a favor by assembling and clearing the land, saving government the time and inconvenience of doing so. However, the land was heavily mortgaged, as the developers did not use their own money. Holders of the mortgages will demand "fair market value." Therefore, Downtown–Uptown's failure is not a windfall to the City of West Palm Beach. If the city had cleared the land with Community Development Block Grant funds, it would have at least controlled what might go on it. Now they must once again respond to the plans of whoever purchases the properties.

7 Alachua County: Running To or From Managed Growth?

Introduction

The previous cases were focused on the Tampa Bay and West Palm Beach areas, where development was initiated after 1890 with the arrival of the railroads. Growth was especially rapid during the period of 1960–1990. Hillsborough County's population increased by almost 440,000 residents during this time, and Palm Beach County experienced a 278.7% increase during the 1960–1990 period, as 635,000 net new residents made that county their home.

In order to gain a broader picture of growth management in Florida, one must examine the dynamics of planning in the older northern part of this state. Gainesville, home of the University of Florida, is located 120 miles north of Tampa, a convenient 2-hour drive up I-75. Gainesville and Alachua County's development history goes back to before the Civil War, as settlement there began as early as 1821, when the United States acquired Florida. This area consists of gently rolling green hills covered by pine and oak forests, as contrasted with the flat sandy landscape of southern Florida. Compared with southeast Florida, the climate is cooler in winter and slightly warmer in summer, as the Gulf and Ocean are about a 1-hour drive in either direction. Alachua County is less crowded and much less dense than Tampa Bay or the southeast Florida megalopolis. Gainesville and Alachua County have experienced much more modest growth than the southern part of the state. Gainesville's population was 29,701 in 1960 and grew to 84,770 in 1990, mainly through annexation. Alachua County grew from 74,000 to 181,596 during this same time, a percentage increase of 145.3%. While this was certainly above the national average for counties over 100,000 population, this is minuscule when compared with Lee, Pasco, and Collier Counties, where the 1960–1990 growth rates were 515.3%, 666.3%, and 863.4%, respectively (see Table 2).

Given a relatively slow growth rate, with accompanying less pressure on the part of development interests to build than that experienced in the three counties mentioned above, one might expect to find a greater level of concern on the part of local governmental and civic leaders for planned growth or even no growth due to conservative attitudes in this traditional Old South area. Another factor that could lead to a controlled growth posture is the presence of the University of Florida, with its small but well-organized group of environmentalists on the faculty and staff. In the late 1960s and early 1970s, it was this group, led by professor Archie F. Carr and his wife Marjorie, that turned the tide against the Cross Florida Barge Canal project when everyone else involved thought that its construction was a fait accompli.[1] Luther Carter notes the presence of the University of Florida during that period of heightened environmental awareness:

> The campaign to save the Oklawaha began in Gainesville, a city some 20 miles from the Oklawaha, mostly among people associated with the University of Florida. In fact, it is fair to say that without this University, there probably would have been no campaign at all; or at least not an effective one. The University with its old buildings, live oaks, palms, and small ponds or sinkholes has the distinctive flavor of North Central Florida....This University, along with Florida State University, is preeminent in Florida, both in the extent of its intellectual resources and in the position and influence of its alumni...its faculty is relatively liberal and includes some social activists.[2]

Given the resources and environmental track record of the University of Florida, along with a lower level of growth pressure than experienced elsewhere in the state, was Alachua County able to use these advantages to prepare the best plans possible under the 1975 and 1985 mandated state regulation? Does the presence of the university — one that established a fully accredited graduate program in urban and regional planning in 1976 — help spur the local planning agency and its staff to be the best they can be? This case examines that phenomenon by means of a microanalysis of a zoning change and macroanalysis in terms of the adoption of the County's comprehensive plan and its subsequent approval by the state.

Background

During the period of 1821–1850, North Central Florida began the process of settlement. By 1860 there were some 8,000 residents in Alachua County, including 3,000 slaves working on several plantations growing high-quality sea island cotton.[3] Gainesville began as the community of "Hogtown" in the 1830s, and in 1854 it became the county seat of Alachua and received its present name in honor of General Edmund P. Gaines, a noted Indian fighter.[4] In 1861, the Fernandina–Cedar Key railroad reached Gainesville, and at that time, this community, platted on the banks of a Hogtown Creek tributary, boasted 958 residents.[5]

After Reconstruction, Gainesville continued to develop as an agricultural trading center for North Central Florida. The freed slaves either worked for their old plantation owners as sharecroppers and/or saved their money, bought land, and became independent farmers, developing their own communities in places such as Rutledge and Jonesville.[6] By 1900, Gainesville had a population of 3,633 according to the U.S. Census. In 1906, the University of Florida was established just outside the city limits. The city grew with the University. In 1910 Gainesville's population was 6,138 and the University enrolled fewer than 300 students. By 1930 there were 10,465 residents in Gainesville and 2,000 university students. By 1950, the city's population increased to 26,861, with 10,000 students.

By 1980, Gainesville had 81,371 residents, and 151,348 people lived in Alachua County. The University of Florida enrolled over 40,000 students served by 3,400 faculty and some 3,000 support staff. The University's medical center, which included a teaching hospital, a Veterans Administration Hospital, and colleges of medicine, dentistry, veterinary medicine, nursing, pharmacy, and health related professions, served the entire 12 county area of North Central Florida,[7] with a 1990 population of 467,972. In the late 1960s Interstate Highway 75 was completed, linking the midwestern United States and Canada with Tampa Bay. Running right past the western edge of Gainesville, this highway generated development of the Oaks Mall, a large regional shopping center of 1.1 million square feet that opened in the late 1970s. The mall served not only north Central Florida residents but also tourists traveling along I-75 who needed food, gasoline, and supplies on their way to and from Disney World and the beaches. This mall now opened the area west of I-75 along Newberry Road to growth and development (see Figure 14).

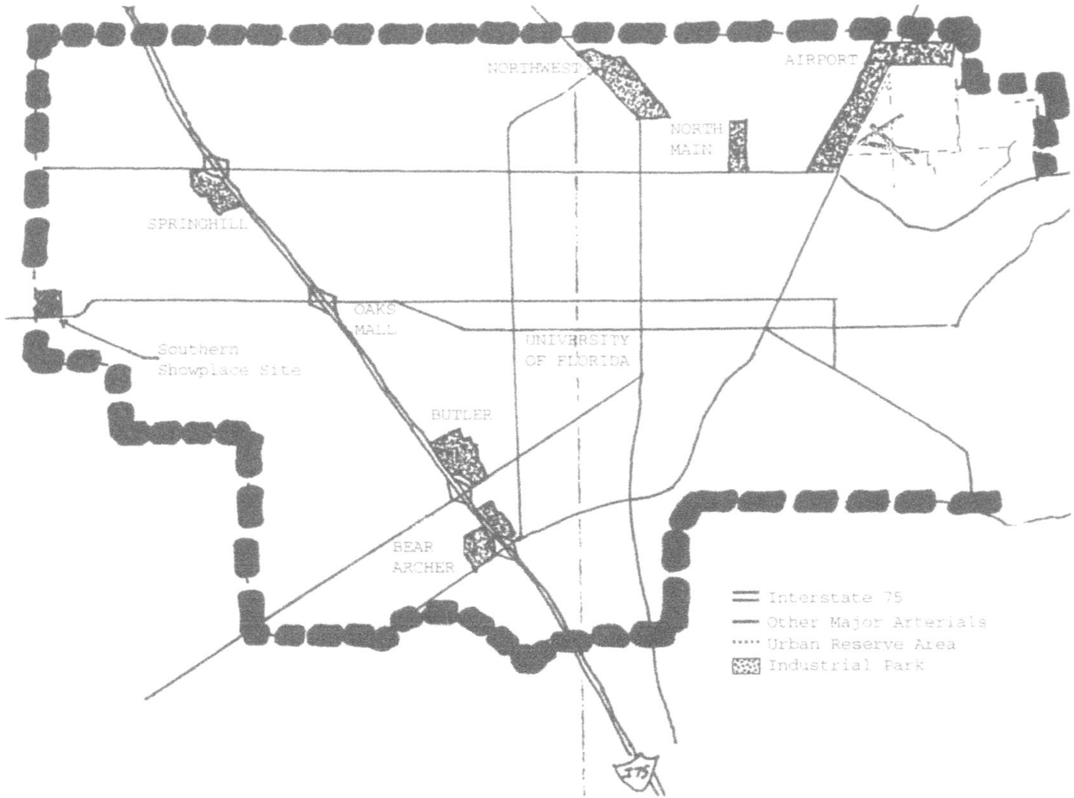

Figure 14. *Industrial Park in the Gainesville Urban Reserve Area, 1990. Source: Alachua County Planning Department. "Survey of Industrial Parks, 1988" (map prepared by author).*

In 1984, Alachua County adopted a plan in accordance with the requirements of the 1975 Local Government Comprehensive Planning Act. The plan stressed protection of Paynes Prairie, an environmental treasure located on the county's southern end, as well as a number of lakes and watercourses. For other lands deemed suitable for urban development, the plan proposed a density of one dwelling unit per acre for virtually the entire county, with higher densities for unincorporated areas on Gainesville's fringe that had already been developed or given subdivision approval. The plan also called for a number of "activity" centers spread throughout the county, usually at the intersection of major roads. These centers were loosely defined, and just like the "CBD" designation in the West Palm Beach City Center Case, could have a variety of land uses, including residential, commercial, office, and industrial.[8]

In 1983, the county adopted a development timing system using "points" for infrastructure provision, such as widened highways and water and sewer lines. The more "points" a developer could claim, the more square footage could be built.

After passage of the 1985 growth management legislation, Alachua County was slow to react and amend its comprehensive plan to comply with the new act's requirements. By December 1, 1990 — although over 260 units of local government had submitted completed plans to the Department of Community Affairs for approval — Alachua County's planners were still compiling data. By the end of 1990, completion of the new plan was not expected until the following spring.

The Southern Showplace Rezoning

In 1973, a group of local investors known as Karma Land Venture, Inc. purchased a 92.79-acre parcel on Newberry Road — Gainesville's major east–west arterial — about five miles west of I-75 in the community of Jonesville. Zoned agricultural, the land remained in pasture and hay cultivation. An equestrian center known as the Southern Showplace opened on the site in 1985 and functioned as a horse and dog show arena, with bleachers seating about 500 spectators and a group of temporary buildings taking up less than 10% of the land area.

After the Oaks Mall opened, residential subdivision activity began in the area on either side of Newberry Road, and by 1990 over 1,000 lots for single-family detached houses had been given subdivision approval within a 2-mile radius of the Southern Showplace site. Several subdivisions were constructed between 1978 and 1990 (see Figure 15) with approximately 400 units built and occupied.

On December 11, 1990, a resident of the newly constructed Jockey Club subdivision in Jonesville noticed a group of small orange signs on Newberry Road on the Southern Showplace site. Inspecting these signs, he found that a hearing for a proposed rezoning of 66 acres to industrial use and the remainder to commercial was scheduled for December 13th by the Planning Commission, just two days away. This resident felt, after assessing the situation, that there was no need to worry: the planning staff would, without a doubt, recommend against it, but it would be a good idea to get as many residents as possible to attend the meeting, because the petitioner will probably be well-prepared and the staff would need support from concerned citizens.

Given the Alachua County's Comprehensive Plan, there was no way the staff could recommend in favor of industrial use on the site. It had been a long-standing professional planning practice to locate industrial uses as close as possible to major transportation arteries to maximize convenience and speed in the delivery of raw materials, shipment of finished products, and movement of workers, and at the same time prevent the intrusion of transport vehicles through residential areas. Examples of this long-standing philosophy were Ebenezer Howard's Garden City (1898), Le Corbusier's Cité Industriel (1925), Ludwig Hilberseimer's "tree theory" (1954), Frank Lloyd Wright's Broad Acres and James Rouse's design for Columbia (1963). There were already three industrial parks in the greater Gainesville area, all located adjacent to I-75 interchanges (see Figure 14). The Southern Showplace site was five miles from the I-75 interchange at Newberry Road, and even from an economic perspective it would be at a competitive disadvantage with the sites on I-75, which had plenty of vacant space.

Another factor was that the site was on the area of high aquifer recharge with sandy soil and limestone subsurface (see Figure 16). A draft of the county's Hazardous Materials Code, dated November 6, 1990, stated with respect to several uses possible in the industrial zone requested by the petitioners "these new storage facilities pose a significant risk of contamination to the Floridan Aquifer and would be located in the portion of

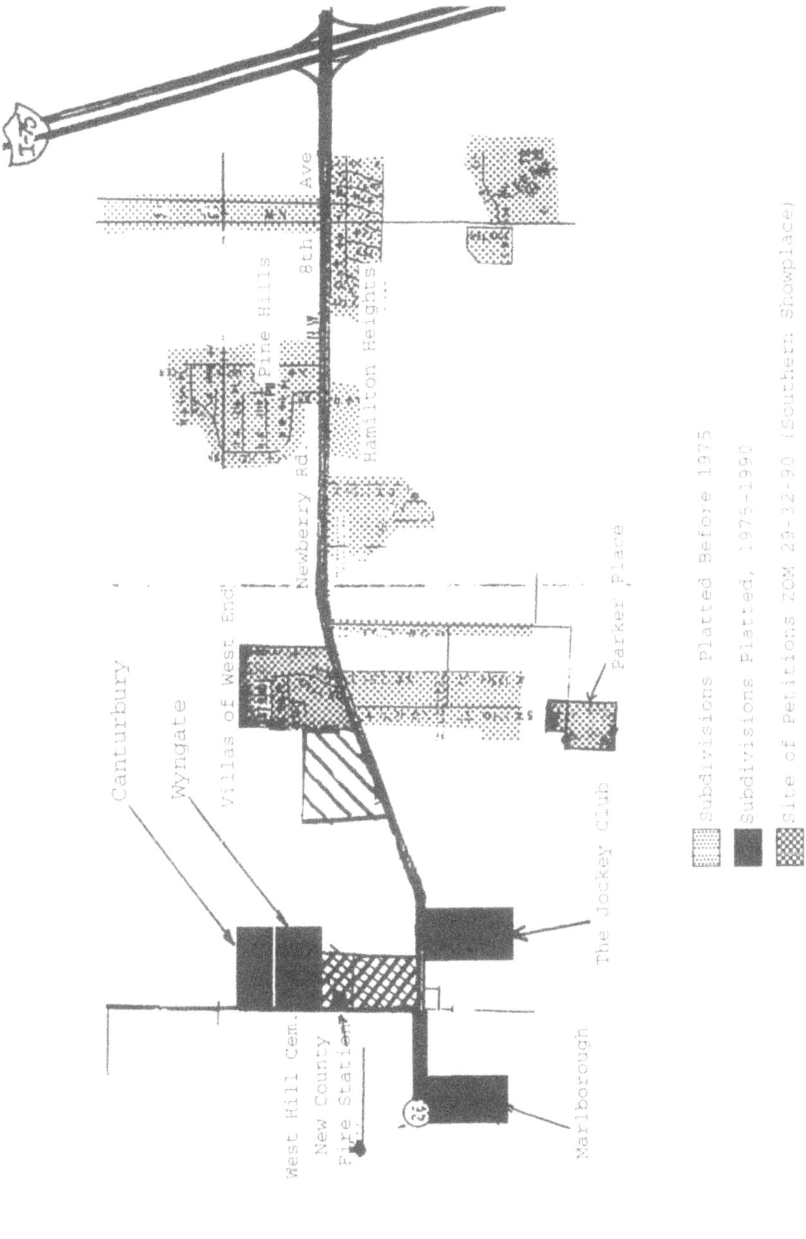

Figure 15. Subdivision activity in Jonesville, 1975–1990.

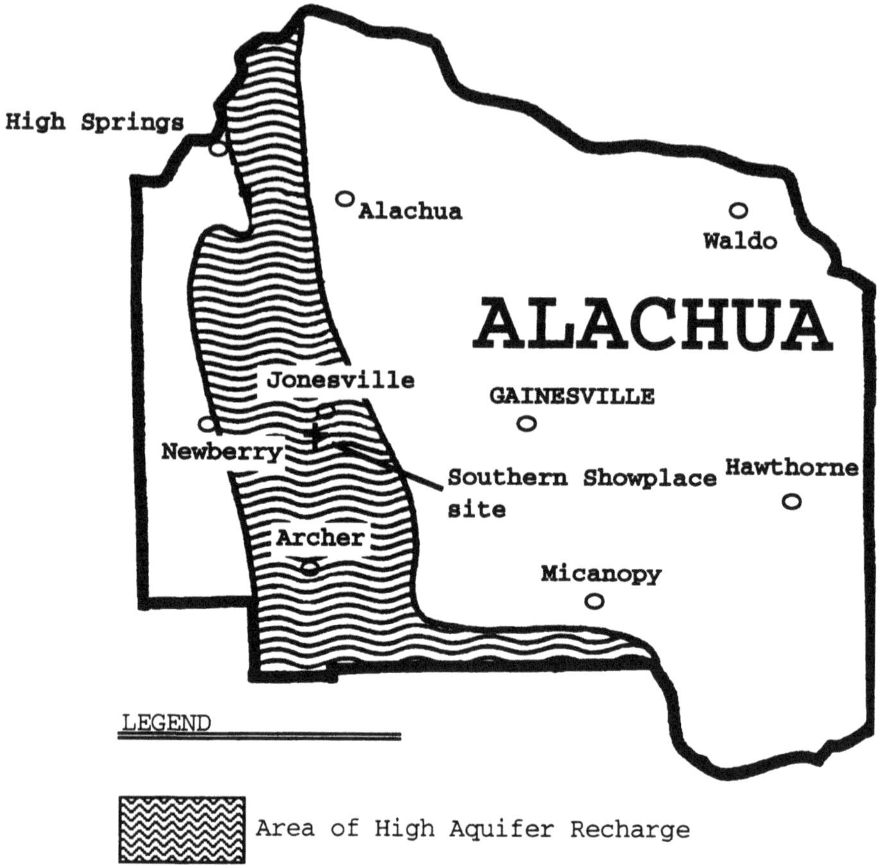

Figure 16. Aquifer recharge in Alachua County. Source: U.S. Geological Survey, 1984.

Alachua County where the Floridan Aquifer is in the confined zone" (out of the high recharge area).[9]

Another factor that mitigated against a professional planning staff recommending industry on that site was that, according to 1990 traffic counts on Newberry Road in the site's vicinity, there were over 12,000 average daily trips (ADT). This meant that, given its two lanes of travel, the road was operating at level "D".[10] Even though this road segment was planned to be widened to four lanes in 1995–96 according to the Florida Department of Transportation's work program, the industrial rezoning, once fully developed, would generate an additional 14,000 ADTs. Coupled with those generated by new subdivision construction already approved, this would bring the total to 28–35,000 ADTs on this road, which translates into a service level "E" or "F."

Confidence in the planning staff went out the window when the next day the staff report recommended approval of industrial zoning.[11] The report was filled with ambiguities. The traffic analysis, while conceding the fact that Newberry Road was operating at level "D," ignored it by stating, *"Rural minor arterials, the functional classification of this road segment are allowed to maintain this level of service and are not considered backlogged."*[12] The fact ignored by this statement was that, with the increasing subdivision activity — and especially if the industrial rezoning was granted — the area would no longer be rural and the much stricter urban standard would apply. While noting that the site was in the unconfined zone above the Floridan Aquifer (see Figure 16), it still recommended approval of rezoning to an industrial classification that could include these uses:

> manufacturing, processing, storage, warehousing, wholesaling, and distribution, ...automotive and heavy equipment sales and repair...public transportation terminals including commercial bus, railroad, air, and water facilities[13]

There would always be a danger that liquid contaminant materials stored in a use permitted in this zoning district would leak into the Floridan Aquifer, but the staff planners ignored this in their report. Also, the staff report only looked at an area 1/4 mile from the center of the rezoning site. This met the legal requirement. However, a rezoning of that magnitude affected residents from at least two miles away.

The staff based their recommendation on a segment of the 1984 comprehensive plan that stated:

> Rural employment centers (of which the site was part of) are characterized by at least one employer of 100 or more persons outside of an urban cluster. These centers may also contain supporting commercial or industrial activities and a concentration of residential development.[14]

The key word is *supporting*. In 1982, Environmental Science and Engineering, a research and development firm, was given permission to build an office and materials testing facility about 1/4 mile south of Newberry

Road. At the time it was constructed based on commercial zoning, the area was indeed rural, but by 1990, subdivisions had marched west from I-75 to within 500 feet of that site. But the rezoning applicant was not Environmental Science and Engineering, but Karma Land Venture, Inc., and their site was over 1,000 feet north of Environmental Science and Engineering, separated from that firm by Newberry Road and a group of retail trade use known as the "Jonesville Mall" (see Figure 17). *This rezoning wasn't supporting anything, except perhaps the profit motives of the Karma group.* Another problem was that the 1984 Comprehensive Plan was prepared when the area around the Showplace site was rural. By 1990, though, it was suburban, and the notion of a Rural Employment Center no longer made any real sense.

The public hearing on this rezoning was held before the Alachua County Planning Commission on December 13, 1990. In the audience were about a dozen residents from West End, a retirement-oriented village 1/2 mile from the site, and five residents of the Jockey Club subdivision located 1/8 of a mile from the Southern Showplace site. There were about 15–20 other individuals in the audience present to testify on other proposed rezonings. The Southern Showplace rezonings were listed as numbers 7–10 on the 20-item agenda, but immediately upon the start of the meeting the Commission Chair moved these items to the end of the agenda.

The meeting began at 7:30 P.M. By midnight, the Showplace rezonings were still to be heard and the only people remaining were the Jockey Club homeowners and a university physics professor who was head of the local Sierra Club chapter. The West End residents, all retirees, had left at 11:15 to go home and go to bed. Finally, at 12:40 A.M., the Showplace Rezonings came up. The petitioner, Karma Land Venture, Inc. presented their side, the staff presented their report recommending approval of the industrial rezoning, and then the Jockey Club residents spoke. All were professionals who had moved to Gainesville due to corporate relocation and were brand new to North Central Florida. They pleaded with the Planning Commission to protect their area "from becoming another Miami."

Other residents spoke, stating their objection to the industrial rezoning as (a) not in concert with accepted professional land use planning practice: trucks traveling to and from the Showplace site would travel through established residential districts; (b) the danger to the Floridan

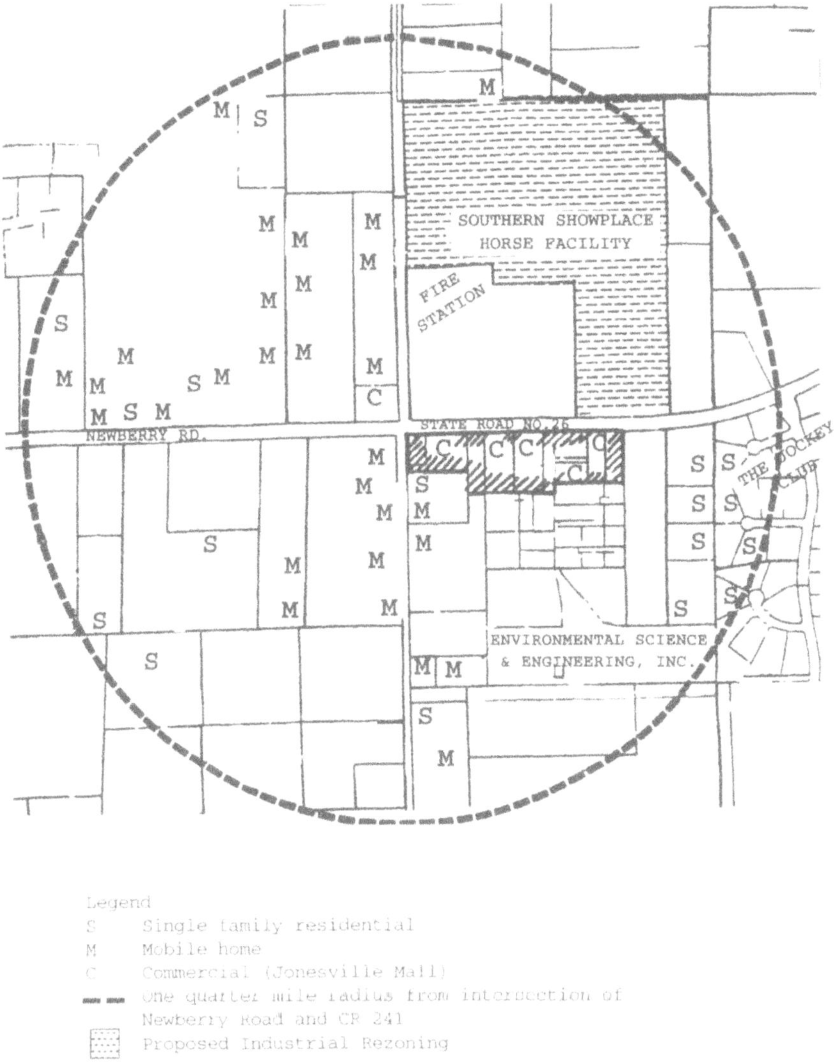

Figure 17. The Southern Showplace site: existing land use, 1990.

Aquifer; (c) a violation of the 1984 Comprehensive Plan, noting that the rezoning did not in any way *support* existing research and development activities of Environmental Science and Engineering; (d) negative traffic impact; and (e) comments from the City of Gainesville's Planning Department that recommended against the proposed rezoning. It seemed unnecessary to rule on a rezoning of this scale at a time when the new

Alachua County Comprehensive Plan was under preparation by the staff and would be released for public review in March 1991. "Why not a moratorium on actions of this type until the new plan is out and approved by your body, the County Commission and the Florida Department of Community Affairs?" the residents proposed in closing. Their remarks were seconded by the Sierra Club leader, who went into considerable detail on the environmental problems presented by the proposed rezoning.

Two of the five planning commissioners present seemed impressed by the Jockey Club residents' arguments — the other three said little or nothing. After discussion and a mild debate between 1:15 and 1:30 A.M., it was felt that the rezoning would be denied. Then, quick as a flash, one of the three "silent" commissioners motioned for approval and it was rapidly seconded. After protests by the two commissioners who seemed to favor the arguments made by the physics professor and the residents, a vote was taken. It was three in favor of the rezoning, two against. The two commissioners who voted against the rezoning, a pharmacist and an administrator in one of the university's support units, seemed embarrassed, but the vote still held. So, at 1:40 A.M., in the early hours of December 14, 1990, one found out what planning in Alachua County was all about.

The Jockey Club Residents Counterattack

Once approved by the Planning Commission, this rezoning was set to go before the Board of County Commissioners on January 8, 1991. The Jockey Club residents were bitter at the response of the planning staff and a majority of the Planning Commissioners, and they joked that the whole thing was set up so that, with the Christmas holidays coming, residents' attention would be diverted and the rezoning would speed through without delay.

Because the decision wasn't made until the wee hours of the morning, there was no local news coverage of the Planning Commission meeting except for those rezonings held before 10:00 P.M., at which time the reporters left.

Some Jockey Club residents made appointments to personally visit with County Commissioners and express their concerns. This tactic ran into a stone wall when one resident asked a Commissioner, "Which one of you represents us? Who is our commissioner?" and was told "We all do!" One of the problems with Alachua County's at-large system — where all five Commissioners are elected countywide — is that no individual Commissioner can be fingered for responsibility on a given site-specific issue such as a proposed rezoning. In this case, the five Alachua County Commissioners represented everyone, but in reality they were accountable to no one, at least not to individual citizens.

The January 8, 1991 meeting was attended by a standing room only crowd in the commission chambers. The County Commission consisted of George Dekle, Chairman, a small business owner; Penny Wheat, Vice Chairman, a divorced, single parent who had the lowest net worth of all five commissioners; Thomas Coward, a realtor and insurance executive; Leveda Brown, a social worker by background and a director of Barnett Bank, one of the two largest banks in Florida; and Kate Barnes, a civic activist who successfully fought to have a neighborhood plan prepared for her area of residence in order to protect the environment and keep out intruding industrial uses. With three women and one African–American male among its five members, the County Commission was indeed a diverse group, but what would their stance on planning be? After visiting with all five Commissioners during late December and early January, residents of the Jockey Club concluded that the only one favorable to their concerns was Penny Wheat, who had a long-term track record favoring planned growth. Some Jockey Club residents had researched records of election campaign contributions for all five Commissioners since 1980 and found that only Penny Wheat had no campaign contributions from local or external development interests.

The meeting proceeded with routine business, and just before the Showcase rezoning came up, the petitioner from Karma Land Venture approached the Commission Chairman and whispered something in his ear. The Chairman called a brief recess, and when the board reconvened a few minutes later, Chairman Dekle stated that the Showplace rezonings would be put off until the next Commission meeting on January 28th. The motion was passed immediately and the residents left knowing they needed to be well prepared in two weeks because Karma certainly would be.

Over the next few days, it became increasingly clear that the Jockey Club residents were up against a powerful juggernaut. Among the Karma Land Venture Inc. limited partners were some of the area's most powerful and influential people. The most prominent person in that investor group was a former speaker of the Florida House of Representatives who now headed a University of Florida research center. The residents talked with the president of the Gainesville Association of Realtors and told her that her group ought to oppose this rezoning because with 18-wheeler trucks running up and down Newberry Road to and from industry on the Showplace site, property values would decline, and her realtors wouldn't be selling many houses. The residents were told, "I can't do anything, but you have to get as many people there as possible and scream your heads off." The entire local development community had closed ranks on this issue even if it meant potential losses to some of their members, namely the local realtors.

Meanwhile, Jockey Club residents found a County Planning Department report on industrial parks. The three parks adjacent to I-75 interchanges were 76%, 71%, and 87% vacant. They were zoned industrial, utilities were available, and they were located in the right places. *No new industrial rezoning was necessary* (see Table 5 and Figure 14).

On January 28th, the Commission Chambers were once again filled to capacity. This time, Jockey Club residents were joined by several elderly African-American farmers who lived in the neighborhood. They were descendants of freedmen who had purchased land there after the Civil War and turned the fields into profitable subsistence farms. While many of these farm families had been bought out over the years, several remained, and these residents were also concerned about the possible intrusion of industrial uses into their community.

At 2:00 P.M., the Showplace rezonings came before the commission. The procedure was the same as that before the Planning Commission: an initial presentation by the petitioner and his consultants, the staff report recommending approval, and then a public hearing where residents were limited to three minutes each, despite the fact that both the petitioner and the planning staff had taken over one-half hour each to make their presentations. Resident after resident spoke against the rezoning. The overriding theme was: "These investors knew the zoning was agricultural when they bought the property. The County Commission owes them no favors." The Commission was reminded that the character of the area had drastically changed since 1984, and therefore a special area study as provided for by the 1984 plan should be undertaken prior to any rezoning action.[15]

Table 5. Analysis of Industrial Parks in Gainesville's Urban Reserve Area, 1988. ***Source:*** **Alachua County Planning Department. Survey of Industrial Parks, 1988.**

Name	Location	Total	Number Vacant	% Vacant	I-75	U.S. Highway	Rail	% in 100-Year Floodplain
Airport	NE 39th Ave Waldo Rd Airport	980	420	42.9%		•	40%	
Northwest	U.S. 441 NW 6th St NW 34th St	790	580	73.4%		•	•	60%
North Main	39th Ave NW Main 53rd Ave	300	280	93.3%			50%	
Bear Archery (SWest)	SW 34th St I-75, SR 121	390	297	76.1%	•			30%
Butler	SW 34th St I-75, SW 34 St	160	115	71.8%	•			0%
Springhill	NW 39th Ave	150	130	86.6%	•			0%
County Totals		2,770	1,822	65.8%				

Within a matter of weeks, the new County Comprehensive Plan would be presented — one that upon approval would guide growth for the next 20 years — through the year 2011. However, the basis for this rezoning — as noted by the County Attorney, the planning staff, and the petitioner — was the 1984 Comprehensive Plan. This plan was now seven years old, and it was terribly outdated, as it failed to anticipate the urban growth west of I-75 that took place during the 1980s. The designation of the Showplace site as a "Rural Employment Center" didn't even make sense in 1984, as that location was only five miles west of the huge Oaks Mall regional shopping center, and the area around this "rural" center was urbanizing as that plan was being written and printed. It seemed to make just too much sense to wait until the new plan was finalized, adopted by the County Commission, and approved by Florida DCA — a process which could be completed as early as June 1991, providing the plan met the test of R9j5. There was no need to hurry, as no one from the petitioner's team was a representative from a corporation wishing to build on the site.

Kate Barnes then spoke for the County Commission: "It seems like we have a group of unhappy residents here tonight. Why don't the planning staff meet with these people for the next month or so and reach a compromise." The other Commissioners, not wishing to make a decision at 11:00 P.M. in front of a capacity audience including newspaper reporters and television cameras, thought this was a good idea. The hearing adjourned and the county Planning Director announced that the first meeting would be held on January 30 in the same County Commission chambers.

The Planning Staff and Citizens Forge a Compromise

Two days later, representatives from the planning staff met with about 50 area residents in the County Commission chambers. The staff began by trotting out a "new" concept of a mixed office and industrial park. It was clear that the staff, for reasons unknown, was trying to get residents to accept a so-called "business park," with offices and just a teeny weeny bit of industry. No one in the audience was buying, though. A Jockey Club resident asked the planning director, whose official title was Director of Growth Management, "Where in Florida have these business parks been tried and how well have they worked?" The "growth managed" director flew into a rage and shouted, "Man, I'll show you some business parks. Come to Dayton, Ohio, come to Irvine. I'll show you some." The staff's credibility with neighborhood residents flew out of the chamber with wings on after that outburst. As the meeting closed, it was clear to the planners that area residents might accept an office development and definitely a neighborhood shopping center on that site, but never, ever, anything with industrial use.

Meanwhile, Karma Land Venture had another card up their sleeve. They presented an amendment to the 1984 Comprehensive Plan that would redesignate the Rural Activity/Employment Center, including the Southern Showplace site, to a "Medium Activity Center." Such a designation would permit an 83-acre business-industrial park, a 210,000–360,000 square foot commercial center, and high/medium-density housing for up to 30,000 people. No economic analysis was submitted to support such a proposal. Existing population within 1-1/2 miles of the site's center was estimated by the planning staff in their report on this amendment at only 1,008, and was projected to grow to no more than 3,000 by 2011 — only 10% of the population Karma was planning for. In a private conversation, planning staffers stated that the Karma proposal was "Field of Dreams" planning. It was like "plan it and it will come." The initial staff report recommended denial, with an alternative plan that called for a center with 200,000 square feet of commercial space with office-industrial uses developed in a "park-like" setting.[16] Given the earlier statements by the staff planners, this alternative seemed like "Field of Dreams II."

On February 7th and 21st, additional meetings were held between the planning staff and neighborhood residents with Karma Land Venture, Inc. representatives in attendance. Jockey Club residents proposed that a maximum of 100,000 square feet for a neighborhood shopping center be provided, including the existing 30,000 square feet of retail use known as the Jonesville Mall. To Karma's credit, that group had donated a small part of the Showplace site to the county for use as a fire station, and had paid for improvements to Newberry Road just opposite their site. The residents understood that Karma deserved *something* for their investment and public improvements — the question was *how much?* By the end of the February 21st meeting, a consensus had been reached between the planning staff and neighborhood residents, who refused to be fooled by the "business park" propaganda.

The staff then revised their alternative proposal for the plan amendment as follows:

1. The activity center be changed from "rural" to a Low Activity Center/Employment.
2. A maximum of 100,000 square feet for a new neighborhood shopping center with development timed to the widening of Newberry Road. An additional 100,000 square feet of commercial use could be permitted once Newberry Road was 4-laned, the population within 1-1/2 miles of the site reached 5,000, and total full-time employment within 1-1/2 miles of the site reached 1,500.
3. An office park would be permitted on the 66 acres previously requested for industrial rezoning by Karma. No industrial uses of any kind would be allowed, and the staff report specifically stated "uses within an office/business park shall not generate noise, lighting, or any other affect that would be a nuisance to surrounding residential development." The office park would be bordered by a 50-foot landscaped buffer to separate it from adjacent uses. A specific list of permitted activities for the office park was added to keep Karma from sneaking in any industrial land use at a later date.

This seemed to be a reasonable compromise. A planning purist would have preferred a 100,000 square foot cap on commercial space, including the 30,000 existing footage, with the remainder of the property devoted to a mix of residential land uses and open space. But, knowing that one doesn't live in a world that is of his/her own brand of perfection, all residents in attendance endorsed the compromise. The Karma people walked out when it was apparent that the meeting was not going their way. They would be back in full force when the planning staff presented the compromise to the planning board on February 27, 1991.

The Planning Commission Overturns the Compromise

On February 27, 1991 the Alachua County Planning Commission met in a public hearing to review the staff compromise report on the Karma plan amendment. Karma's representatives went first, requesting approval of the amendment as initially submitted, with industrial uses as part of the development. This time they added a new wrinkle: the industrial uses would produce jobs. This was an important issue in Gainesville, as wages are depressed due to the presence of college students willing to work, either full- or part-time, for low pay. Alachua County has virtually no industry, and the 1980 U.S. Census showed that among 323 metropolitan areas in the United States, Gainesville ranked 303 in median household income. The only problem with Karma's approach was that at no time since the December 13th meeting when this process began was there an industrialist speaking as part of the Karma group pleading for the rezoning in order to build an establishment that would create jobs.

The pharmacist, who favored planned growth, and was so amenable to neighborhood concerns at the December 13th meeting, had resigned from the Commission in disgust. His replacement was an African–American parole administrator for the State of Florida who had publicly gone on record as supporting anything that brought jobs to Gainesville, which was understandable, as 90% of his caseload, mostly African-American men, were unemployed at the time they had committed their crimes. While watching the petitioners with their newly found interest in jobs for the underclass, one couldn't help but think that they were simply playing to the new Planning Commissioner's feelings.

After Karma Land Venture, Inc.'s presentation, the staff presented their compromise report, forged one week before. The presentation was weak and without conviction. It seemed that the staff presenters were in a hurry to finish and then run out of the room. Several residents spoke on behalf of the compromise. One spoke on the issue of industrial use in the area, with all of the negatives, also mentioning the "psychology of industry," i.e., the notion that people are uncomfortable with living next to industry unless it is separated by major highways or rail lines, such as provided by the time-tested concepts of Planning Theory and Practice, and this discomfort would result in a lowered tax base due to decreased property values. One resident concluded by saying that "granting an amendment permitting more square footage in commercial office and industrial use than could ever be absorbed by the market will have a chilling effect on the entire movement toward planned growth, as evidenced by the process of give and take between the planning staff, residents, and the developer that we have undergone over the past month."

The Planning Commission Chairman, John Don Puckett, an architect by profession, replied, "If we don't give the petitioners what they want, that will have a chilling effect on development." Puckett then ordered the planning staff to meet with the developers (Karma) and return in one week with a new recommendation. The staff meekly agreed and one week later returned to the planning board with a new list of possible uses, including warehousing, storage, and materials assembly. The maximum permitted square footage for commercial use was increased to 200,000, without constraints such as population size and number of employees. The "new" staff proposal was adopted 6–1, the only dissenting vote coming from the same research center administrator who voted against the industrial rezoning proposal for the same site on December 13th.

The meeting before the County Commission on the plan amendment was set for March 18, 1991. By this time, residents were thoroughly confused as the planning staff began to release drafts of the new Comprehensive Plan in early February. Now people were wondering whether the March 18th meeting was on the Southern Showplace zone change, the 1984 Comprehensive Plan amendment for the same site, or the new future land use plan, which proposed another "activity center," again for, the same site. Also, neighborhood residents were disheartened over the apparent doublecross by the planning staff.

The March 18th meeting was on the plan amendment. Before a standing room only audience and the local television station cameras, one resident, Steve Humphrey, a University of Florida research biologist, spoke with eloquence and conviction. Looking directly at Commissioner Kate Barnes, he said, almost shouting, "You told us to meet with the planners and reach a compromise. We did just that. Now the L.P.A. [Planning Commission] rejects our compromise and we're looking at the same old industry none of us wants." Other residents echoed the same theme. The planning staff stated over and over that their "new" staff report, created by the demands of the Planning Commission Chairman, was virtually no different than the compromise. When residents objected, the County Commission chair ruled them over their allotted time to speak and out of order. Finally, at 2:15 A.M., after five hours of discussion, the County Commission agreed to accept the planning board recommendation of 200,000 square feet of commercial space plus a "business park," including industrial uses, on the 96-acre Southern Showplace site in the form of a "Low Medium Activity Center."

With a major victory in hand, the Karma Land Venture Inc. group then hired a landscape architect to prepare a planned unit development business park design with industrial and office uses for submission to the Planning Commission and then the County Commission in lieu of the proposed 66-acre industrial rezoning. An attractive site plan was prepared, and the proposal sailed through Planning Commission review without opposition. The County Commission hearing on the rezoning was scheduled for June 25, 1991.

On June 24th, an article appeared in the *Gainesville Sun* entitled "Tiny Critter: Big Impact."[17] University of Florida naturalists had found that just under the Showplace site was the nesting place of the 1-1/2-inch long squirrel chimney cave shrimp, a rare species. Steve Humphrey was quoted as saying:

> The only place in the world it (the cave shrimp) is known to occur is downstream as the water flows. Any pollutants that enter the groundwater, flow in that direction. It is conceivable that if there is a spill, it could cause that animal to go extinct. By all of these same facts, that same spill could cause our drinking water to become polluted.

The same news article quoted the general partner for Karma Land Venture and the leading presenter for that group at all of the public meetings as saying:

> The existing Alachua County industrial zoning is by and large irrelevant and unusable in the marketplace. It is in wetland areas and considered by and large unstable. And it's on the eastern side of Alachua County and not acceptable to industry because most now require their sites to be near the interstate.[18]

This person was referring to the airport, Northwest, and North Main sites that indeed have portions of their sites in wetlands ranging from 40% to 60% (see Table 6). He conveniently forgot to mention the Bear Archery, Springhill, and Butler industrial park sites located directly adjacent to I-75 interchanges; 71–87% vacant, with only one having any wetland area (Bear Archery, 30%).[19]

With the cave shrimp article, the Jockey Club residents now had the smoking gun and free publicity that had been lacking for so long. However, by now, the area residents were tired and confused with the rezoning, plan amendment, and the new comprehensive processes all swimming around in their heads. Neighborhood interest and enthusiasm had peaked

Table 6. Growth in North Central Florida 1980–1990.

County	1980 Population	1990 Population	% Change
Alachua	151,348	181,596	20.0
Baker	15,289	18,486	20.9
Bradford	20,023	22,515	11.2
Columbia	35,399	42,613	20.4
Dixie	7,751	10,585	36.6
Hamilton	8,767	10,930	24.7
Lafayette	4,035	5,578	27.7
Levy	19,870	25,923	30.5
Marion	122,488	194,833	59.1
Putnam	50,549	65,070	28.1
Suwanee	22,287	26,780	20.4
Union	10,166	10,252	0.9
Total	467,972	615,161	32.00%

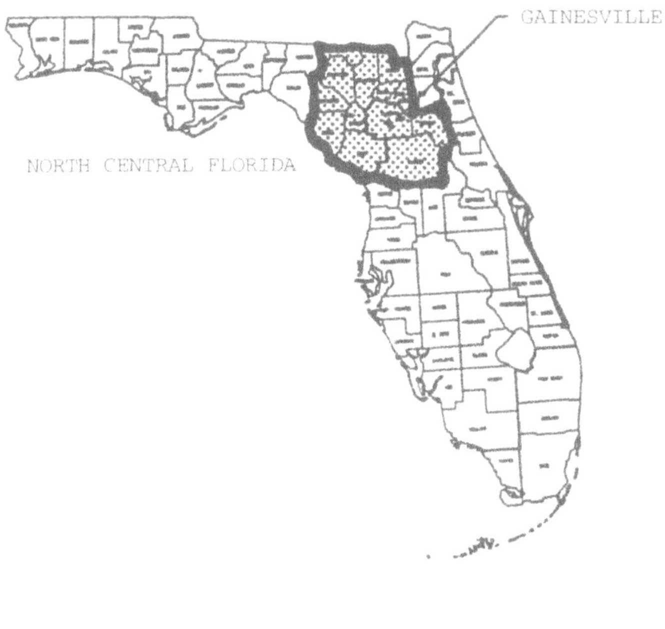

at the March 18, 1991 meeting, when the fight for the compromise plan amendment was lost. This was very unfortunate, because the rezoning before the County Commission on June 25th was really the important event, while the March 18 meeting, as well as all previous ones, was simply preliminary. The issue on the table was rezoning from Agriculture ("A") to Planned Unit Development (PUD), which was for the same 66-acre portion of the site as the December 13th rezoning from Agriculture to MS (Local Service Industrial or Light Industry). What had occurred was that the petitioner had simply withdrawn the original proposal, and then after the confusion over the plan amendment to the old 1984 plan and the revised plan itself had set in, replaced it with the new rezoning scheme.

At the June 25th meeting, the Commission chamber was less than half full. Present were Steve Humphrey and a few diehard neighborhood residents, including some of the black farmers who had attended every previous meeting since late January, but never spoken. Karma presented their site plan, which was more definitive than the black-bordered rezoning drawing shown in January. Staff planners now contended that the new County Hazardous Materials code would protect the cave shrimp and the Floridan Aquifer. But that argument was in conflict with the code's language as stated earlier.[20]

Only three County Commissioners attended this final meeting on the Southern Showplace site rezoning: Kate Barnes, Chairman George Dekle and Thomas Coward. Leveda Brown was absent, as was Penny Wheat, the people's champion and the only unabashed supporter of planned growth among the group. The residents thought out loud, "Where is Penny Wheat when we need her?"

The residents asked that a vote as important as this be put off until the full Commission was present to hear all sides once and for all. Kate Barnes seemed to tune in to this theme, but the other two Commissioners pressed for a vote. At 11:30 P.M., the Commission suspended discussion on this rezoning, moving on to other business. By 2:35 A.M., after almost everyone in the audience had left to go home, the rezoning was placed back on the agenda and quickly voted on. By unanimous decision, the rezoning from Agriculture to Planned Unit Development was approved.

The area residents now organized as the Greater Jonesville Neighborhood Association, with Stephen Humphrey as President, and filed suit in August 1991 to overturn the County Commission's decision.[21] After months of legal maneuvering, the suit was finally withdrawn on January 28, 1993 when the Association ran out of money. The residents then turned their attention to political action to try and vote out of office as many Commissioners as possible who were beholden to developers and opposed to planned growth.

The Alachua County Comprehensive Plan of 1991: DCA Says NO!

On May 1, 1991, Alachua County Commissioners sent the revised comprehensive plan to the Florida Department of Community Affairs (DCA) for review as per the 1985 Act's requirements. On August 9, 1991, Robert G. Nave, Director of the Division of Resource Planning and Management, Department of Community Affairs, sent to County Commission Chair George Dekle a 102-page statement strongly critical of the County's plan.

That statement was known as "Objections, Recommendations, and Comments."[22]

The primary concern, as stated in Nave's letter, referred to the plan's future land use elements map or FLUM. Quoting from the letter:

> Our review revealed several concerns with your proposed comprehensive plan that we consider critical. These include, but are not limited to: the FLUM is inconsistent with plan objectives and policies to direct development to appropriate areas and protect natural resources; the FLUM designates residential use for areas which are not projected to develop during the planning period; inadequate policy language to ensure the protection of natural resources (lack of specific guidelines and standards); and an inadequate concurrency management system.

The report contained a total of 234 specific objections and comments; while perhaps 80% were technical and procedural in nature, the major one was about allowable residential densities. The county's plan called for a one-dwelling unit per five acres in the rural areas, while the Department of Community Affairs called for either a cap on the number of development permits that could be issued for zoning forests and other areas least likely to be developed at lower densities, such as one unit per 20 acres, as was the case for several "urban" counties with plans previously approved by DCA (see Figure 18).

The most significant objections noted by DCA were:

1. The method for calculating needed development is professionally unacceptable for the allocation of residential, commercial, and industrial development sites;
2. The plan promotes urban sprawl, including leapfrog development;

3. The plan allows owners of undeveloped land to build under existing but inconsistent zoning for up to two years after adoption of the plan (vesting);
4. No protection is provided for groundwater recharge areas;
5. Other natural resources suitable for conservation are not adequately identified or protected;
6. Numerous parts of the plan are not supported by data and analysis showing a need for development;
7. Goals do not state long-term ends toward which development is directed;
8. Objectives are vague and not measurable;
9. Policies are tentative, conditional or do not describe the activities and regulations for implementing the plan;
10. The maps do not accurately show existing or proposed land uses and facilities for public services (traffic, drinking water, sewage, solid waste);
11. Needs, timing, location, and costs of capital improvements are not identified; and
12. The plan is not consistent with the state comprehensive plan or the regional policy plan.

Public reaction to the DCA report was mixed. Conservationists felt that Alachua County's government should comply with DCA's requirements. Iris Burke, a University of Florida Law School professor and board member of Friends of Alachua County, was quoted as saying "Comprehensive planning isn't just a good idea. It's the law. And the purpose of the law is to protect our economy and our environment by planning wisely for the future."[23] However, Howard Wallace, a member of the Alachua County Chamber of Commerce's Planning Task Force, was quoted in the same news article as: "They (DCA) want to tie our hands and tell us what density to accept. I don't think that is what the legislature really wanted. I don't think that's what the people of Florida want."[24]

A series of public workshops were scheduled with the County Commission and the Planning Commission during the month of September. The first workshop was to be held on September 1, at 7:30 P.M. Area residents called the planning department several times before the meeting to request copies of the proposed revisions prepared by the staff to satisfy, at least in part, the concerns of DCA. Everyone was told, "We don't have any yet." One minute after the meeting started, a planning staffer

County	Lowest Gross Density	Cap	Area Allocation	Cluster	Overplay	Performance Standards	Point System	Rural Village	Transfer Development Rights	Family Farm
Flagler	1DU/20AC			X	X	X		X		1DU/5AC
Hillsborough	1DU/20AC			X		X		X		1DU/5AC
Bay	1DU/10AC			X	X	X				No
St. Johns	1DU/10AC	X		X		X	X		LDR	DU on same parcel
St. Lucie	1DU/40			X					LDR	
Martin	•	X				X				No
Duval	1DU/100AC		X							•
Jefferson	1DU/20AC			X				X		Yes
Dixie	1DU/40AC									No
Polk	•						X			•

Figure 18. *Rural Planning Techniques of Other Florida Counties. Source: Abstract of information provided by the Department of Community Affairs and the Florida Association of Counties on stipulated settlements and compliance agreements.*

distributed copies of the revision to the audience. After the staff presentation, public discussion began. Speaker after speaker criticized the planning staff for not having material made available at least a day or two beforehand. A representative from 1,000 Friends of Florida, the environmental group that years before led the fight against the Cross-Florida Barge Canal project, stated that even with the revisions the plan violated state law; the proposed land uses were not based on suitability analysis, which is a tried and true means of determining appropriate land use; and that the plan lacked vision.[25]

On September 9, 1991, another public workshop was held, with members of the Planning Commission and the County Commission attending. The attitude of the Planning Commissioners and some of the County Commissioners was: "DCA can go to hell. We're going to do our own thing." Several representatives of development interest testified, urging that the county maintain the one dwelling unit per five-acre land use designation in the rural areas. The county's growth management director replied that by clustering proposed development such a density could be maintained and at the same time comply with DCA's preference for an overall one dwelling per twenty acres limit. When County Commissioner Penny Wheat asked, "What locations would be appropriate for this type of use?" the growth management director had no answer. One of DCA's objections was that the plan did not provide a map showing the aquifer recharge areas. This was important, because if such a map was included it would show that certain areas were not appropriate for certain types of development, such as industrial uses on the Southern Showplace site in Jonesville. Commissioner Wheat asked the growth management director "Where is the aquifer recharge map?" The director replied, "We have no maps like that." I whispered to Steve Humphry sitting beside me, "Every planning department in Florida has these maps and a couple of staff planners told me the maps were in the office." Humphrey smiled and added, "If he doesn't have them the U.S. Department of Agriculture certainly does." County Commissioner Kate Barnes then entered the discussion, stating that use of a development cap recommended by DCA would be "unfair to property owners."

The official public hearing on the planning staff's response to DCA was on October 1, 1991. The County Commission chamber was filled to capacity, with the audience evenly divided between development interests and environmentalists. The planning staff agreed to supply all of the necessary maps and tabular data requested by DCA and proposed a cap of 100 dwelling units per year for 20 years in those parts of the county deemed rural (i.e., outside of the urban reserve area). However, the planning staff, after being urged by a majority of the Plan Commission members to do so, refused to budge from their initial rural density of one dwelling per five acres. As per the 1985 Act's provisions, opponents of the plan established standing to speak at any future administrative hearings between DCA and Alachua County by testifying against the proposed revisions or submitting a letter to the meeting reporter. After a discussion that lasted until 1:45 A.M., the county commissioners agreed to submit the staff's proposal to the Florida Department of Community Affairs as an official response to the "Objections, Recommendations, and Comments" statement issued six weeks beforehand.

On November 24, 1991, a large, ominous legal notice appeared in the *Gainesville Sun.* The notice, officially entitled *"State of Florida Department of Community Affairs Notice of Intent to Find the Alachua County Comprehensive Plan Not in Compliance, Docket NO 91-no1-0101-(N)"* was a clear signal that DCA was unhappy with the planning staff's response transmitted as per the October 2, 1991 vote of the Alachua County Commission (see Figure 19). On November 25, 1991, DCA issued a statement of intent to find the plan in noncompliance.[26] DCA's concerns were stated in five areas:

1. Land uses are not backed up by adequate and relevant data and analysis that show the county can support that many new dwelling units or that much new industry without creating urban sprawl.
2. Urban housing densities are allowed up to a half mile outside of the urban area without a change in the plan.
3. Language on the protection of natural resources such as wetlands, aquifer recharge areas, and wildlife habitats is vague.
4. Land use densities are not adjusted to protect natural resources such as wetlands and aquifer recharge areas.
5. The plan allows logging within conservation areas without establishing standards to ensure protection of natural resources.

Despite DCA's objections, that department opted to take a conciliatory stance. Tony Arrant, a DCA planning manager assigned to work with Alachua County, stated, "These are serious issues. But they are not issues

STATE OF FLORIDA
DEPARTMENT OF COMMUNITY AFFAIRS
NOTICE OF INTENT TO FIND THE ALACHUA COUNTY COMPREHENSIVE PLAN NOT IN COMPLIANCE
DOCKET NO. 91-N01-0101-(N)

The Department gives notice of its intent to find the Alachua County Comprehensive Plan, NOT IN COMPLIANCE pursuant to Section 163.3184, F.S.

The adopted Alachua County Comprehensive Plan, the Department's Objections, Recommendations and Comments Report, and the Department's Statement of Intent to Find the Comprehensive Plan Not in Compliance will be available for public inspection Monday through Friday, except for legal holidays, during normal business hours, at the Alachua County Office of Planning and Development, 10 Southwest Second Avenue, Third Floor, Gainesville, Florida 32601-6294.

This Notice of Intent and the Statement of Intent will be forwarded to the Division of Administrative Hearings of the Department of Administration for the scheduling of an administrative hearing pursuant to Section 120.57, F.S. The purpose of the administrative hearing will be to present evidence and testimony and forward a recommended order to the Administration Commission.

An affected person, as defined in Section 163.3184, F.S., may petition for leave to intervene in the proceeding. A petition for intervention must be filed at least five (5) days before the date set for the final hearing and must include all of the information and contents described in Rule 221-6.010, F.A.C. A petition for leave to intervene shall be filed at the Division of Administrative Hearings, Department of Administration, 1230 Apalachee Parkway, Tallahassee, Florida 32399-1550. Failure to petition to intervene within the allowed time frame constitutes waiver of any right such person has to request a hearing under Section 120.57, F.S.

/s/ Robert Pennock, Chief
Department of Community Affairs
Division of Resource Planning and Management
Bureau of Local Planning
2740 Centerview Drive
Tallahassee, Florida 32399-2100

Figure 19. *Notice of intent to find Alachua County* not *in compliance.*

the DCA has not seen before and they are not issues we haven't seen settled."[27] Arrant went on to say that about 50% of local plans fail to comply when they are returned to DCA after the "Objections, Recommendations, and Comments" statement is issued. DCA staffers indicated that an administrative hearing could not be held until June 1992 and that meanwhile county officials and DCA staff would try to negotiate a compromise.

While individuals who have established standing as indicated beforehand can testify in administrative hearings there is no provision for citizen participation in "negotiations" between DCA and local government officials. Vincent A. Cautero, the Director for the Development Services Department for Citrus County Florida, spoke succinctly on this phenomenon:

> At the negotiation meetings, policies, data, analysis, and criteria in the plan are discussed along with ways to bring the insufficient portions of the plan into compliance. The public may attend, but not speak. For all practical purposes, it is a closed public hearing. This is a mixed blessing. The process is beneficial as it allows professionals from both sides to delve into issues and talk technically without unnecessary deliberation by the public. It is frustrating for the public to be able to listen but not speak. According to interpretations of the Florida Sunshine Law, State officials will not deny citizens access to a meeting in a State building in which a plan in their community is discussed. The staff members from the community's legal and planning departments cannot talk openly about the "politics" of their plan due to the fear that excerpts from these discussions will be brought up on the record at future public hearings in the community.
>
> The public never had the chance to oppose a change until a formal public hearing. Thus, a deal could be cut by state and local staff before a public hearing is ever conducted.[28]

In describing a case study in Citrus County, Cautero shows that when DCA found that county's plan in noncompliance in June 1989, a negotiated settlement was reached in December of that year; but after the agreement was signed, property owners, and private groups opposed the terms. The Citrus County Board of Commissioners approached DCA to renegotiate and DCA agreed. New terms were agreed to in May 1990, and while the amendments became final in November, neither the Board,

Builders Association, Board of Realtors, or the Chamber of Commerce were happy.[29]

DCA Now Says YES!

On January 21, 1993, the settlement between the Florida Department of Community Affairs (DCA) and Alachua County Government became final. DCA conceded on virtually every point they raised in the "Objections, Recommendations, and Comments" statement of August 9, 1991. *The maximum density for rural land use remained at one dwelling per five acres, as originally proposed by the County planning staff.* Minor adjustments were made to the proposed activity centers in terms of timing development with population growth. The technical objections relating to language, required data, maps, etc., made earlier by DCA, were removed after some minor rewriting by the planning staff.[30] Still, the most important point remained that DCA backed off of its most significant objection — that of maximum dwelling unit densities for rural land use. Now, developers will be able to sell 5-acre "ranchettes" all over the county except in defined protected areas such as wetlands.

Just as in the Citrus County case documented by Cautero, all parties were not happy with the settlement. The Jonesville Activity Center, as designated by the draft compromise plan, called for a "business park" on the Southern Showplace site, with permitted uses including "light assembly of products from previously manufactured material." In a hearing on November 23, 1992, the Greater Jonesville Neighborhood Association successfully argued for removal of the industrial category from permitted uses for the Jonesville Business Park based on the potential of such uses to contaminate groundwater. A majority of the Alachua County Commissioners directed staff to bring back language with industrial uses in Jonesville deleted.[31]

The planning staff then proposed language that was extremely restrictive. Their new proposal removed industrial uses from the entire Activity Center, and it prevented environmental consulting companies as well. This would have prevented Environmental Sciences and Engineering from any expansion, so that now the staff was pitting the Association against a major employer. Additionally, the proposed language applied the Hazardous Management Activity codes to all uses in the Business Park.

This would have prevented establishments from storing or selling common items such as fertilizer, chlorine for swimming pools, and chemicals for household use.[32] The Association argued at the final public hearing on January 14, 1993 that the staff proposal was a trick to convince the Commissioners to retreat from their former position. After angry denials from the planning staff, the Alachua County Commission voted 3 to 2 to retain the industrial uses as first proposed by the state. This recommendation was then incorporated into the final settlement agreement with the Florida Department of Community Affairs.[33]

Summary

At the beginning of this chapter we posed the question that, given (1) a growth rate considerably lower than elsewhere in the state, producing in turn lower levels of pressure for development, and (2) the presence of the University of Florida, the state's premier institution of higher education, would elected and appointed officials and civic leaders be more amenable to managed growth? In recent years, given the Southern Showplace rezoning and the preparation of the 1991 Comprehensive Plan, that answer may be a somewhat guarded "perhaps not." The county leadership's attitude toward planning is probably best expressed by a letter to the editor of the *Gainesville Sun* written by Lou Ann Callahan and printed on August 24, 1991. Entitled "County Plan a Developer's Dream," the letter read as follows:

> I have followed the development of the proposed Alachua County Comprehensive Plan with great interest. My faith in democratic government was thoroughly tested when, during the public meetings, our elected officials and employees hired by those same officials, in my opinion, gave preferential treatment to the developers.
>
> Public input was overshadowed by the developers' show of force with paid "experts." The process never demonstrated a balance about a common goal, but pitted developers and policy-makers against the general public.
>
> I had the distinct impression that decisions were made with arrogance and the attitude of "we know what we are doing."
>
> With the report from the Department of Community Affairs in Tallahassee, our officials received their reply this month. Though the

> report only deals with basic policies, it is obvious that Alachua County did not pass the grade.
>
> The DCA stated a number of objections in the report. Parts of the report read: "...are not measurable or specific and are not supported by adequate and relevant data and analysis." The report also says that "a number of goals, objectives, policies, standards, findings, and conclusions within the proposed comprehensive plan are not supported by relevant and appropriate data...contained vague conditional language...and is not based upon a professional acceptable methodology."
>
> This is not complimentary. As a matter of fact, the DCA's response points out the unprofessional attempt at trying to pass off a comprehensive plan that is not comprehensive at all.
>
> That which should serve to guide Alachua County's growth management and lay a foundation for our county's future was nothing more than a developer's dream come true, to be manipulated and permitting speculation through vague language and ambiguities.
>
> In *The Gainesville Sun* editorial of August 15, the last paragraph reads "Alachua County blew it in the 1970s during Florida's earlier attempt at growth management."
>
> This comprehensive plan is for 1991 to 2011. We cannot afford to blow this one, too. How long can we hide behind the learning process? The county has a professional planning staff. It is time they show professionalism and stop favoring developers. After all, do not taxpayers pay for their salaries?
>
> Our elected officials also need to better represent the residents of this county by demonstrating to the public that Florida is not for sale. The same officials should not become developers' putty, but vote without bias on the issues brought before them with a view to the future and quality of life for the residents of this county.

That writer could have added that not only were decisions made with "arrogance," but always after 1:00 A.M., which had the effect of eliminating news coverage in the papers the following day.

Ms. Callahan's frustration is supported by my observation that four of the five County Commissioners and six of the seven Planning Commission members seem by their actions in the issues cited in this case study to be development-oriented and disdainful of planning intruding on business as usual. The county's planning staff, realizing that their superior officers are pro-development, fall right in line because, after all, they have mortgages and auto loans to pay and they want to send their children to college. So even in the face of conduct that borders on violations of the American Institute of Certified Planners (AICP) code of ethics, they do what they have to do in today's recessionary times with a limited job market for professionally trained planners.

In Tampa Bay, West Palm Beach, Miami, and Jacksonville, planners are faced with very real large-scale development proposals. These include the planned communities of Tampa Palms, Wellington (West Palm Beach) and Weston (Broward County), stadiums such as the Florida Suncoast Dome in St. Petersburg, or Joe Robbie on the Dade-Broward County line, entertainment centers such as Disney World and its satellites, and mega "suburban downtowns" or "edge cities" such as Cypress Creek in Broward County or the Avenue Mall and related office development in South Jacksonville. These projects are planned, DRIs are filed if necessary, and they are built. Jobs are created and the property generates higher tax yields than before. In Gainesville, projects of this type and magnitude don't happen. In the last 20 years the only construction project that even remotely approached the scale of those mentioned above was the Oaks Mall, built in the 1970s.

Instead, the development game in Alachua County is land speculation, not construction. The Southern Showplace rezoning, one of the largest in this county proposed over the past 10 years, was not initiated by a corporation wishing to build something there now. After all, the location, 5 miles from an interstate highway interchange, was not competitive for industrial or office construction with industrially zoned vacant land sites directly off I-75 and adjacent to its interchanges. However, if the rezoning had been granted, as originally planned, Karma would have the option to sell to a long-term investor at a higher price because land zoned for industry is worth much more than if zoned for agriculture. Property owners in rural areas were aghast at a plan with densities at 1 dwelling per 20 acres, because only with 1 per 5 acres could they print attractive brochures advertising 5-acre "ranchettes" in a pastoral wooded setting and successfully market the vacant land to empty-nester couples on the verge of retirement and fed up with the cold and decay of northern U.S. cities or the congestion and fear of crime in Southeast Florida.

In Chapter 3, I mentioned some potential problems with the 1985 Local Government Comprehensive Planning and Land Development Regulation Act (LGCPLDRA). These included questions as to (1) whether

R9j5 was really necessary, given the increased number of certified planning professionals in the state between 1975 and 1985; (2) lack of funding for planning by either the state or federal government; (3) the watering down of requirements for consistency between local plans and those of the regional policy plans and the state comprehensive plan by the so-called "glitch" bill; and (4) whether the "top-down" approach is reasonable in terms of its ability to cut through 120 years of laissez-faire state governmental attitudes toward growth. This case study uncovers two additional possible problems. First of all, a moratorium should have been placed on zone changes while work on comprehensive plans was in progress. Not having a moratorium placed the Alachua County government in the embarrassing position of having to rule on a major zone change using a 7-year-old comprehensive plan that was out of date the day after it was printed. At the same time the zone change was being reviewed, a new plan was in the final stages of preparation. Such a moratorium need not be placed on zoning variance or conditional use permits; only large zone changes need to be included. We can define "large" as a zone change application for sites larger than one acre per 10,000 residents in the governmental jurisdiction involved. Given Alachua County's 1980 population size of just over 150,000 residents, only zone changes for sites of 15 acres or more would be affected by moratoria.[34] Another problem is vesting. The framers of the 1985 act should have learned from the failures of the 1975 version and eliminated vesting completely. Once the plan is approved by DCA for those holding property that is inconsistently zoned, unless they have a building permit in hand they would be required to build according to plan specifications.

It seems that given the experiences in Alachua County and that documented by Vincent A. Cautero for Citrus County, the DCA negotiation process posed serious problems for citizen participation and the concept of open government. Were DCA review and approval as structured in the early 1990s *really* necessary given the fact that almost all of the plans were rejected but DCA wound up negotiating away at least half of their objections? Is negotiation really just a form of administrative plea bargaining that allows the guilty to go either unpunished or under-punished? Were the DCA reviewers professionals and full members of the American Institute of Certified Planners (AICP) or simply clerks trying to follow written guidelines prepared by others? In the American Institute of Certified

Planners' 1990–91 roster only five members of this organization were reported to work for the Department of Community Affairs. Were there enough fully trained planners to handle all 461 units of Florida local government?

The role of the University of Florida (UF) with respect to the quality of planning in Alachua County is interesting. Since 1976, the university has maintained a fully accredited department of urban and regional planning where 12 full-time faculty and research scientists — all of whom had either their Ph.D., AICP membership, or both — teach and conduct research in a graduate level program that offers the master of science in city and regional planning. While several City of Gainesville planners are graduates of this program, and faculty routinely serve on the City's Planning Board, there have been no faculty serving as members of the County Planning Board (Local Planning Agency, or LPA) between 1987 and 1992, and before 1993 no graduates of the UF planning program were on the paid professional staff. County government does not routinely seek help from the UF planning department — not even the free assistance in the form of student projects provided by most other universities across the nation with graduate programs in urban and regional planning. It seems as though, when planning is concerned, the influence of the University of Florida ends at the Gainesville city limits, which are no more than eight miles in either direction from the center of UF's campus.

Perhaps one reason for this rather unfortunate situation is that actually the Gainesville area has a dual structure of governance — one for local government, another for the university. Political, business, and civic leaders in Gainesville, and more so, Alachua and the other 11 counties that make up North Central Florida, tend to be locally born and bred and may or may not have UF ties in terms of formal education. However, university personnel from the president to untenured assistant professors are almost always outsiders raised and educated elsewhere. The university establishment has from the start been viewed with hostility and distrust on the part of key locals. The hostility is masked with almost unbelievable politeness, as in this truly southern region of the nation strangers speak to each other on the street, and every adult is nicely addressed as "sir" or "ma'am." But if "outsiders" intrude on the native turf — especially if they get involved in controversial issues such as planning vs. unlimited growth — then the welcome mat is quickly pulled from beneath their feet.

The norm in North Central Florida is not the openness and freedom of thought one tends to associate with university communities. While there are now over 6,000 faculty and professional employees at the university, North Central Florida's economic base is really the two dozen state correctional, mental health, and mental retardation facilities, which employ a total of just over 10,000 workers. This creates a "corrections" environment, with the top of the pyramid being the electric chair at Florida State Prison, 20 miles from the University of Florida. The prevailing mood is a tight, iron-fisted authoritarianism thinly veiled by the velvet glove of southern hospitality. As far as planning is concerned, it manifested itself in public meetings on the Southern Showplace rezoning and plan amendment, where citizens were arbitrarily cut off by the County Commission Chairman after *he* decided that their time was up. In my years as a professional urban planner working as a practitioner consultant or researcher in dozens of cities, including Tampa, West Palm Beach, Hollywood, Orlando, and Eustis, Florida, I have never seen citizens treated as shabbily by elected and appointed officials as they were in Alachua County. And after the arrogant and cavalier treatment, decisions were made between midnight and dawn with virtually no one present except the Commissioners themselves and county staff.

Conclusion

Michael Peter Smith found that during the late 1980s grassroots movements of the middle class in Sunbelt cities would soon have the effect of policy alterations and even restructuring.[35] Both the microexamination of planning policy as shown by the Southern Showplace rezoning issue and a more macro-examination of the evolution of Alachua County's 1991 Comprehensive Plan tend to reinforce Smith's point.

In the Showplace rezoning, Karma apparently thought no one would see the orange signs and they would get their blanket industrial rezoning in place *before* the comprehensive plan was complete and then gain vesting for at least two years. When the issue was finally resolved they had to prepare a planned unit development, which was much more restrictive than a simple rezoning. They also wound up with the Greater Jonesville Neighborhood Association being created as a result of the controversy. That organization will now, and figuratively forever, be looking over their shoulder.

Despite the shortcomings of the Florida Department of Community Affairs (DCA) review process, particularly the negotiations that produced a settlement, the plan that resulted was much better than the one initially submitted to DCA in terms of environmental protection. More important, the plan review process aroused many of the conservationist groups in Alachua county which had been dormant since the environmental movement locally, statewide, and nationally crested in the mid-1970s.

Although one might postulate that half a loaf is better than none at all, this is not the point. Given the resources of the University of Florida, coupled with a need to protect the beautiful rolling green landscape of Alachua County covered with pine and oak with an occasional palm tree here and there, much more could have been done in both micro and macroscale. Hopefully, despite the problems and mistakes of the past, it is not too late to craft a brighter future for this uniquely beautiful part of Florida.

Notes

1. Luther Carter. *The Florida Experience: Land and Water Policy in a Growth State,* Johns Hopkins Press (Baltimore), 1975, p. 279.
2. Ibid., pp. 278, 279.
3. Charles H. Hildreth and Merlins G. Cox. *History of Gainesville and Alachua County,* Alachua County Historical Society (Gainesville), 1981, pp. 1–23.
4. F.W. Bucholtz. *History of Alachua County,* The Record Press (St. Augustine), 1920, p. 6–8.
5. Jess G. Davis. *History of Alachua County,* Alachua County Historical Society (Gainesville), 1954, p. 3.
6. Ibid., pp. 127, 146.
7. North Central Florida includes Alachua, Baker, Bradford, Columbia, Dixie, Hamilton, Lafayette, Levy, Marion, Putnam, Suwanee, and Union Counties. Population was 467,972 in 1980 and 615,161 in 1990, a 32% increase (see table). The major growth was in Marion County, which increased by almost 60%.
8. Alachua County Comprehensive Plan. Future Land Use Element, Map and Text, 1984.
9. Draft, Alachua County Hazardous Materials Code, November, 6, 1990, amended January 4, 1991.
10. Source: Institute of Transportation Engineers, *Trip Generation Manual,* 4th edition, 1987.

11. Alachua County Planning Department Staff Report on 20M-30-90, dated 12-12-90.
12. Ibid., p. 8.
13. Summary, Zoning Regulations, Alachua County, Florida. Amended through 6-30-90, p. 5.
14. Alachua County Comprehensive Plan, Policy 7.4, *Rural Employment Centers,* 1984.
15. Policy 11.1.1 of the 1984 Comprehensive Plan, *"Special Area Studies,"* states "in those areas undergoing rapid land use transition, special area studies shall be conducted to evaluate the special needs of these locations."
16. Comprehensive Plan Amendment CPA, 1-90. Staff report prepared 1-10-92, p. 16.
17. Kay Stokes. "Tiny Critter, Big Impact: Cave Shrimp Could Prevent Development Stumbling Blocks," *Gainesville Sun,* June 14, 1992, page 1A.
18. Ibid., p. 6A.
19. Alachua County Planning Commission: Inventory of Industrial Parks, 1988.
20. Op. cit. Note 9. The *Gainesville Sun* reported on June 24, 1991 that the Hazardous Materials Code seeks to protect the county's water supply by limiting the sort of development that can occur in areas such as Jonesville, "where the thick underground clay layers that protect the aquifer elsewhere are absent."
21. Kay Stokes. "Jonesville Residents Sue County," *Gainesville Sun,* August 22, 1991, p. 1D.
22. The statement is provided for by Rule 9J-11.010 Florida Administrative Code. Also supporting that statement are § 163.3184 Florida Statutes and Rule 9J-11.011 Florida Administrative Code.
23. Kay Stokes. "Views Split on Comply Plan Critique," *Gainesville Sun,* p. 1B, August 18, 1991.
24. Ibid., p. 2B.
25. Kay Stokes. "Plan's Protection of Nature Lacking," *Gainesville Sun,* September 13, 1991, p. 1B.
26. Kay Stokes. "County Plan Doesn't Meet State Criteria," *Gainesville Sun,* November 26, 1991, p. 1B.
27. Ibid.
28. Vincent A. Cautero. "Growth Management in Florida: The Essence of the Negotiated Settlement," in *Planning and Public Policy,* Vol. 16, No. 1, University of Illinois (Champaign–Urbana), November, 1991, p. 5.
29. Ibid., p. 6.
30. Settlement Agreement, Florida Department of Community Affairs and Alachua County Board of Commissioners January 21, 1993, pp. 5,712,714.

31. James Hellegaard, *Gainesville Sun,* "Don't Forget Us, Jonesville Tells County," November 24, 1992, p. 1b. Also, Varqui Wright, "Attorney Criticizes Comprehensive Plan," *Gainesville Sun,* November 24, 1992, p. 1A.
32. Greater Jonesville Neighborhood Association, Inc. Newsletter No. 8, August 1993, pp. 1, 2.
33. Antonia English, "Land Use Plan Consensus Reached." *Gainesville Sun,* January 15, 1993, p. 1A.
34. If only the unincorporated area of Alachua County is included, that population, which was 70,000 in 1980, would create a threshold for sites seven acres or more.
35. Michael Peter Smith. 1988. *City, State and Market: The Political Economy of Urban Society,* Basil Blackwell Inc., (New York, NY).

8 Toward Improving the Planning/Development Process for Florida's Local Governmental Units

Introduction

Chapters 2 and 3 showed that, for the first 120 years of Florida's history, growth without limits was encouraged by the state's leaders. In the period of 1965–70, many changes took place: reapportionment of the legislature; the 1968 constitution, which permitted governors to succeed themselves; "government in the sunshine"; and a growing environmental movement statewide and nationally all helped to create a mood conducive to planned growth. The election of Reubin Askew as governor in 1970, coupled with a series of environmental crises, built on the previous developments to goad the legislature to enact sweeping, widespread legislation that finally led to the Local Government Comprehensive Planning Act of 1975 and its successor in 1985.

The case studies in Chapters 4–7 were taken from experiences as a result of this legislation in Hillsborough County, Tampa, West Palm Beach, and Gainesville (Alachua County), with references made to the plan approval process in Citrus County. The broad observation from these case studies is that state legislation by itself was no guarantee of quality plans being adopted and implemented. The quality of planning will depend on local commitment to a successful process on the part of elected officials, governmental administrators, local planning staffs, and citizen leaders.

While the state can provide a *framework* for planning, unless local government becomes convinced that planned growth is really an idea whose time has come, the process will be less than successful. To use the old adage once again: *"you can lead a horse to water, but you can't make it drink."*

Given the history of Florida's state and local governmental response to growth and development, coupled with what we have learned from the 1975 and 1985 legislation, how can we better plan for Florida's growth in the years beyond? First, I will present a review of the process leading to passage of the 1993 Planning and Growth Management Act, followed by references to recent citizen surveys on planning. Then, specific recommendations will be made in the areas of (1) changes to the 1993 Planning and Growth Management Act; (2) changes in local governmental structures and elections; and (3) a program of citizen education in urban and regional planning. While these three sets of observation and subsequent recommendations apply directly to Florida, the implications are national in scope.

The 1993 Planning and Growth Management Act

In Chapter 3 we saw that in 1970 the successful candidacies of Reubin Askew for Governor and Lawton Chiles for U.S. Senator lifted Florida from the political dark ages and into the sunlight of moderation. Both Askew and Chiles were young, moderate and pro-business on one hand, but they also understood the very real need to protect Florida's delicate environment. Askew went on to a second term as governor, effectively retired from private life in 1979, and well into the 1990s served as a quiet elder statesman for the Florida and national Democratic party. Lawton Chiles was reelected to the U.S. Senate in 1976 and 1982. In 1988 he decided against reelection, claiming "burnout."[1]

By late 1989, Chiles had a change of heart and announced his candidacy for governor of Florida in the 1990 primary and general elections. He breezed through the primary with minimal opposition and then soundly defeated one-term Republican incumbent Robert "Bob" Martinez. It was Martinez who, in 1987, repealed the service tax that was established in order to, among other things, provide monies for Florida growth management (see Chapter 1). Upon Chiles' election, planners in Florida and elsewhere felt that, given his relatively liberal past and campaign pledges to protect the state's fragile physical environment, Chiles would strengthen growth management in Florida.

However, 1990 was much different from 1970. The national recession which began in 1989 hit Florida very hard in the primary areas of tourism, retirement, and leisure, with immediate aftershocks felt by the land development and construction industries. The watchwords for 1990 and the years to follow were "Economic Development," not the 1970s theme of environmental protection. Also by 1990, problems with sections of the 1985 legislation, as shown by the case studies presented here, were apparent throughout Florida particularly with respect to the effectiveness of Developments of Regional Impact (DRI), R9j5, concurrency provisions, and review of local government plans by the Department of Community Affairs (DCA) for compliance with legislative intent.

In 1991, Governor Chiles convened the third Environmental Land Management Study Committee. Known as ELMS III, this group, as contrasted to the balanced 1972 and 1984 ELMS committees, was heavily tilted toward the interests of developers. Of the 46 members, 13 were developers, 11 were lawyers, most of whom were associated with developers, and 9 were elected officials, all of whom had accepted campaign contributions from development interests. Four were appointed officials, including the Secretary, Department of Community Affairs, the Chief Administrative Officer of the City of Jacksonville, the Broward County Administrator, and the Assistant City Attorney of Coral Springs. There were six "citizen representatives," including the Executive Director of the 1000 Friends of Florida, the Director of the Center for Governmental Responsibility of the University of Florida and a member of the ubiquitous Karma Land Venture Inc., mentioned in the preceding charter.

There were only two professional planners on this body, V. Gail Easley, past president of the Florida Chapter of the American Planning Association, and Richard C. Bernhardt, Director of Planning and Development for the City of Orlando. The lone academic on ELMS III was Dr. John DeGrove, Director of the Joint Center for Environmental and Urban Problems and the author of several major books and articles on Florida growth management.[2]

The ELMS III report, entitled *"Building Successful Communities,"* was completed and transmitted to Governor Chiles on December 15, 1992.[3] The committee, headed by James Harold Thompson of the law firm of Ausley, McMullen, McGehee, Carothen, and Proctor, and Linda Loomis Shelley, Secretary of the Florida Department of Community Affairs, presented what was essentially a blueprint for economic development, as contrasted with previous ELMS findings, which were oriented more toward environmental enhancement. There were 174 recommendations, the most significant one being that "the legislature should provide incentives to encourage local governments to make economic development an integral part of the local planning program."[4] Other major recommendations, including provisions of a new "Strategy Growth and Development Plan" to be placed "strategically" between the State Comprehensive Plan and Local Government Comprehensive Plans; phasing out the Development of Regional Impact (DRI) program; reducing the 26 required subjects in the current Comprehensive Regional Policy Plans to only five in the new proposed Strategic Regional Policy plans; and weakening concurrency by allowing local governments to "establish adequate level of service standards for most state roads without demonstrating consistency with State-set minima as currently required."[5]

ELMS III downgraded concurrency, by stating:

> Concurrency, while it unquestionably serves an important public purpose, may lead to inefficiencies in some development activities. These effects must be minimized.[6]

The ELMS III recommendations were quickly transformed into state law. In the 1993 legislative session, the Planning and Growth Management Act was rushed through both houses and signed by Governor Chiles.[7] Significant changes from the 1985 legislation included:[8]

State Planning. Biennial review of the State Comprehensive Plan: preparation of a "growth management portion" of the State Comprehensive Plan.

Role of the Regional Planning Councils. These bodies were virtually stripped of their regulatory authority and instead downgraded to a focus on planning coordination. They were removed from regulatory authority to appeal development orders for Developments of Regional Impact (DRI). These councils must adopt "Strategic Regional Plans" with only five required subjects — affordable housing, economic development, emergency preparedness, natural resources of regional significance, and regional transportation.

Local Comprehensive Planning. Provisions were made for the intergovernmental coordination element. The intergovernmental coordination element established regulatory processes now served by the DRI program. This element must also direct all agencies to consider the implications of proposed action on private property rights. A new plan amendment process eliminated most Objections, Recommendations, and Comments Reports, which was a significant item of discussion in the previous chapter on Alachua County.

Concurrency. The Act confirmed that concurrency applies as a matter of state law only to potable water, sanitary sewer, drainage, solid waste, park and recreational facilities, roads, and mass transit. It made standards for meeting concurrency requirements "flexible," even allowing municipalities in some cases to set their own standards.

Development of Regional Impact. Once the new intergovernmental coordination elements are in place, the DRI program will be phased out. The date for DRI phaseout is set now for December 31, 1997. The DRI program may remain in certain small and rural governmental areas (counties under 100,000 population) but with increased thresholds for certain land uses in urban downtown and regional activity centers.

Other Provisions. A new local option gas tax of up to five cents per gallon was made available without a need for voter approval. A campus master planning program for all State University System campuses was also provided. This legislation also revised the law on municipal annexation to prevent unincorporated enclaves or "islands." In all, of the 174 ELMS III recommendations, 130 were reflected in the 1993 legislation.[9]

Criticisms of the 1992 ELMS III report and the 1993 Planning and Growth Management Act were quick to follow. In a 1993 issue of *Florida Planning,* Max Forgey, Planning Director for Charlotte County attacked ELMS III for being dominated by development interests and failing to listen to local governmental planners.[10] According to Forgey, ELMS III had subordinated visionary planning to the public administration incrementalist tasks of growth management. At the Seventh Annual Growth Management Conference held in Fort Lauderdale in April 1994, Debbie Orshefsky, an attorney with 14 years of experience with Developments of Regional Impact (DRI) was skeptical about the new rule transferring DRI review to local governments, stating:

> The rule is very vague — and under it, we will have DRIs on acid. Local governments will have to identify developments that will have a significant impact on other communities. But significant impact is not defined. These projects will be subjected to some other type of "development review process" with the applications of mitigation criteria. DRIs won't go away, they'll just be given another name.[11]

The staff of Florida Atlantic University–Florida International University, Joint Center for Environmental and Urban Problems also took a swipe at the new DRI provisions, stating that "How well the act succeeds will depend largely on local governments," and "In short, the amendments to the DRI process have created uncertainty and confusion about how large-scale developments will be addressed in the future."[12] The staff noted with respect to changes in concurrency requirements that local governments could adopt long-term concurrency management systems that would permit new development despite deficient transportation facilities as long as these deficiencies would be corrected within 15 years. The staff also noted that local governments would adopt whatever level of service standards they chose for all roads other than the Florida Interstate System. The staff concluded their remarks on concurrency with the following statement:

> Viewed in a more cynical light, the exceptions to transportation concurrency might be seen as so numerous and open-ended as to undermine the concept of transportation facilities keeping pace with growth. Exceptions for in-fill and redevelopment and special treatment of transportation concurrency management areas can be used to exempt entire cities and long term concurrency management may be used to delay for substantial time periods, transportation concurrency in suburbs.[13]

Given the shift of emphasis from environmental protection to economic development, the deletions of regulatory authority from the Regional Planning Councils, and the weakening of concurrency requirements, this author feels that the 1993 legislation represented a retreat from planned growth compared to the initial 1975 Act and its much stronger 1985 counterpart. As we saw in the Old Hyde Park, West Palm Beach City Center, and Alachua County Cases, 1975 and 1985 DRI regulations were easy enough to get around. With DRI regulation weakened by the 1993 legislation, the opportunity for governmental oversight with respect to large-scale development is virtually abolished.

Florida's Citizens' Attitude Toward Planning and Zoning

In fall 1984, the Florida Atlantic University–Florida International University Joint Center for Environmental and Urban Problems, along with the FAU Social Science Research Laboratory, conducted a series of statewide opinion surveys on citizen attitudes toward environmental issues.[14] The issues included environmental quality, pollution, legislation, protection of fish and wildlife, water conservation, preference for economic growth, and land use regulations. Findings were strongly in favor of environmental protection in all three regions of the state.

Statewide, 53% felt that Florida's environment was getting worse, 83% felt that the state needed stronger laws to prevent pollution, 90% agreed that Florida needed stronger laws to protect fish and wildlife, 83% also disagreed with the notion that water conservation was not needed, and 79% disagreed that economic growth should take place even at the expense of environment. This was an important finding, because the survey was conducted just two years after the 1979–1982 recession was concluded. Central to the scope of this book, *55% reported that land use regulation should be strengthened.* While 24% felt that land use regulation should be left "as is," only 6% were in favor of relaxing land use controls. In the northern region of the state, 51% favored strengthening land use regulation, compared with 58% in the central region and 55% in South Florida. Using statistical analysis, the surveyors found that in the northern and central regions, opposition to economic development is associated with a belief that regulation is needed for water conservation, and support for strengthened land use controls is related to support for legislation for the protection of wildlife and prevention of pollution. In the southern region of Florida residents seemed almost equally concerned about all areas surveyed.[15]

The same study looked at the Treasure Coast region in detail. This region includes Indian River, St. Lucie, Martin, and Palm Beach counties, all of which experienced heavy growth during the period of 1960–1990, and especially since 1980. Of residents surveyed, 57.5% said they wanted stronger land use regulation and only 6.5% advocated relaxing land use controls. Of those who wanted strong land use controls, 46.5% gave as their main reason problems in the planning and zoning process, another 33.9% favored stronger land use controls because of a desire to improve the quality of the built environment, and another 17% said their major reason for strengthening land use controls was to fight environmental problems of water pollution, water shortages, the destruction of wildlife, and loss of agricultural lands.[16]

A similar study among residents of the Florida Keys also identified strong, positive attitudes toward environmental protection, and the need for planned growth. In telephone interviews with 408 residents (95% confidence level and 5% interval of error), 59% felt that the environment was getting worse and 59% felt that land use regulations should be strengthened. Only 13% felt that land use regulations should be relaxed.[17] As for who is not paying their fair share for growth, 53% of those surveyed said "developers," 23% blamed tourists and winter visitors, while 43% fingered state government.

Given these surveys, it seems that a majority of Florida residents uniformly throughout the state favored planned growth. Also, a large majority of elected officials, civic leaders, and development community representatives *publicly* favor some level of planning. How can this announced support for planning be translated into strong and effective implementation programs throughout the state of Florida? Perhaps the following recommendations for change will supply some answers.

Recommendations

These recommendations are based on the literature cited in the first three chapters, the case studies, the mood of Florida's citizens as evidenced by surveys, cited earlier here, and numerous conversations continuously since the late 1970s with local planners and DCA officials on the merits of the 1975, 1985, and 1993 Acts. They are stated in three sections: (a) changes to the 1993 Planning and Growth Management Act; (b) changes in local governmental structure and election process; and (c) a program of citizen education.

Changes to the 1993 Planning and Growth Management Act

Given that 1993 legislation represents a retreat from planned growth, it is important to point out what is *good* about the 1985 legislation. It did correct several problems associated with the 1975 Act, namely quality and uniformity of plans and lack of relationship between state goals, regional policies, and local plans. Its most important contribution, though, is that along with the 1975 Act, it has raised citizen awareness as to the need for planning, as evidenced by the surveys conducted by Florida Atlantic University and referred to earlier in this chapter. Now virtually every one of Florida's 461 units of local government has been through not one, but two cycles of planning. Things have changed drastically since 1977, when in the weeks before adoption of Horizon 2000, Hillsborough County staff planners were replying in response to tough questions by citizens, "Planning is so new to us."

Now that the 1993 Planning and Growth Management Act is with us and will guide growth in Florida well into the next century, it should be basically retained, especially in terms of required elements and consistency with regional and state plans and the Strategic Growth and Development Plan. But it could be improved by the following measures:

1. Addition of a Mid-Range Development or Strategic Plan to the Local Government Comprehensive Plans

Currently, the legislation requires only 20-year long-range comprehensive plans. This creates a gap between day-to-day "immediate actions such as zone change requests and the long range plan." The local government planners are forced to make recommendations on zone changes to appointed and elected officials based on a blob or circle on a map labeled "activity center." Comprehensive plan goals, objectives, and policies that talk about what a community should be like in 20 years and development regulations that guide today's decisions on land use change are not enough to do the job of effective planning. We saw the problem created in the Southern Showplace rezoning by lack of a middle-range plan, and we also saw the advantage of a mid-range scheme in the Old Hyde Park Village case, where the Hyde Park neighborhood plan was detailed enough to give neighborhood residents and government planners a leg to stand on in their negotiation with the developer.

The mid-range development plan is not a new concept. It goes back at least as far as the urban renewal plans of 1949 and Martin Meyerson's classic article in 1956 entitled "Building the Middle Range Bridge for Comprehensive Planning." In discussing the detailed development plan as part of his "bridge," Meyerson stated:

> Long-range comprehensive plans commonly reveal a desired state of affairs. They rarely specify the detailed courses of action needed to

> achieve that desired state. By their long-range nature they cannot do so. The development plan in contrast, will indicate the specific changes in land use programmed for each year, the rate of new growth, the public facilities to be built, the structures to be removed, the private investment required, the extent and sources, of public funds to be raised, the tax and other local incentives to encourage private behavior requisite to the plan.[18]

Almost 30 years later, Frank So refined Meyerson's concept by providing a definition of strategic urban and regional planning as a seven-step process of (a) internal research/analysis; (b) analysis of external factors; (c) analysis of strengths and weaknesses; (d) the implications of a, b and c; (e) development of strategic objectives; (f) implementation of objectives; and (g) monitoring and feedback. According to So, strategic plan objectives must be direct, to the point, and quantifiable, as contrasted to the broad policy recommendation of long-range comprehensive plans.[19] These strategic, mid-range plans would cover a 5- to 10-year period and be prepared on the neighborhood or sub area scale. Elements of mid-range plans could be limited to future land use, circulation, and infrastructure provision. Mid-range plans would be closely linked to the capital improvements element. While these plans would involve considerable expense, by coordinating projects to be built within a long-range *and* mid-range planning framework, waste and duplication of effort would be minimized, and over time these plans could pay for themselves.

2. Eliminate R9j5: Replace It with a Declaration

R9j5 would have made a great deal of sense in 1975, when planning in Florida was in a primitive state. Even in 1985, R9j5 made some sense, as the products of the 1975 act varied so much in terms of quality. However, the number of AICP-certified planners in the state has tripled since the mid-1970s. Most local governments have gone through two complete cycles of planning in less than 20 years, an experience that has given even the most remote hamlet a growing sophistication about planning.

Once all units of local government have fully complied with the 1993 Act's requirement, I would suggest that a new cycle of plan updates be prepared. During this time, mid-range plans could be prepared and the adoption process simplified. In place of R9j5, the only requirement would be a one-page declaration signed by a full member of the American Institute of Certified Planners attesting that the plan meets with acceptable standards of planning practice of the American Institute of Certified Planners. Because of possible legal liability (i.e., suits by business or citizens), no AICP member would sign and notarize such a statement unless the plan did in fact cover all elements, back goals, objectives, and policies with required data — and in sum, do all of the things R9j5 was supposed to do. Such an action would empower the planning profession, eliminate costly review by DCA, and avoid the ominous "negotiation" process, with its built-in problems for open government as noted earlier by Cautero. The Florida Department of Community Affairs could then focus on reviewing plans for consistency with regional and state plans and do this with less staff than at present.

3. *Improve State Funding for Planning*

Lack of state funding was a problem hampering full effectuation of both the 1975 and 1985 Acts. *Adequate funding must be provided in any future round of new plans.* Rather than blanket funding to communities on a per capita basis, a better way would be a weighted formula, with consideration including not only population size, but more so, past performance in planning as measured by timely submission to DCA and subsequent approval of plans from the 1985 and 1993 Acts, the use of optional elements, and extent of sub area planning. Plans that were approved by DCA after only initial objections, recommendations, and comment statements were sent to local government would be first in line for funding. Those governments who went into a negotiating process with DCA would be next, and last in line would be those communities that had to be dragged into administrative hearings by the state. While the 1993 Act made provisions for a local option gas tax, which could be used for plan implementation, local governments are under no obligation to pass such a tax, especially with the 1995 climate of "smaller" government.

Funding on the basis of past performance would reward those communities that took planning seriously and made an effort to comply with the legislative intent of the 1985 LGCPLDRA and the 1993 Act, not so much to satisfy DCA, but because they saw planning as a means of preserving and enhancing their cities and counties. While such a carrot and stick approach would work to the disadvantage of some communities, its initiation would do much to minimize noncompliance during the next round of plan updates.

Instead of relying on a local-option gas tax, a statewide gas tax increment could be collected by the state and redistributed to local governments for concurrency purposes.

B. Changes in Governmental Structure

Most of the growth management literature focuses on the details of planning requirements. Another question that should be asked is, What type of governmental framework best enhances planning?

In the West Palm Beach City Center case, vice-mayor Pat Pepper-Schwab was quoted as saying "We have leaders but they're not providing leadership. We react instead of being leaders, which means being out there in front painting the vision." In the Showplace rezoning in Alachua County, local residents opposed to the rezoning to industrial were frustrated when, while inquiring as to which one of the five county commissioners represented them, were told, "We all do." These two instances, which have been multiplied time and time again in my 20-year experience with planning in Florida, focus on a major problem area. The council-manager form of governance, and legislative body members being elected at large, both of which are products of the early 20th Century reform movement, do not work well for planning because there is a built-in lack of accountability. The council-manager form, with at-large elections, accounts for 92% of Florida's over 460 local governments. To correct the problems presented by that form, I pose two sweeping recommendations:

1. Replace the Council-Manager Form of Government with the Strong Mayor Variety in Cities over 50,000 and Counties over 100,000 Population

In the strong mayor form of government it is clear where the buck stops. Legislative and executive functions are clearly defined and separated, whereas in the council-manager form they are blurred. Collective responsibility means no one is really in charge, certainly not the "mayor," who is really no more than a first among equals. Herbert Smith, in his analysis of planning "success," notes that despite an individualistic political culture, Baltimore and Pittsburgh have done well due to, among other things, a strong mayor form of government, where accountability is centered in a single office.[20]

West Palm Beach has already responded to the lament of Ms. Pepper-Schwab and others. In November 1991, that city of 70,000 residents switched from a council-manager to strong mayor and elected Nancy Graham as the first mayor under that new format. Only time will tell if Ms. Graham effectively managed growth and development, but no one can question whether she had the authority for decision making to accompany responsibility for the city's future.

A strong mayor system need not be a return to the "ward politics" of old. West Palm Beach retained a City Manager to handle the day-to-day responsibility of government. The City Council is still the body responsible for being the final arbiter on zone changes, plan adoptions, and amendments. The mayor, however, is "out front painting the vision" setting the tone for, among other things, the quality of planning.

2. Replace "At-Large" Election with Election by District for Cities with over 50,000 and Counties with More Than 100,000 Residents.

At-large election of the local legislative body (City Council or Commission) has its roots in the Progressive reform movement which swept the nation during 1900–1915. The Progressives thought that if the entire electorate in a given local governmental jurisdiction voted for the legislative body as a unit, elected officials would be concerned with the city or county as a whole, not just the concerns of a segment.

The political science literature since the mid-1970s has been generally critical of at-large elections because of their negative impact on the election of blacks and Hispanics to office. But at-large governance has another effect that relates directly to planning issues: *a collective body elected at-large can avoid accountability to localized groups of individuals, especially in NIMBY cases.*

Members of legislative bodies elected at-large have to campaign successfully throughout the entire jurisdiction in order to win. Executing a large-scale campaign means raising a war chest to buy radio and television time, newspaper advertisements, and the like. It also means gaining newspaper endorsement. In order to "sell" themselves to the voters, candidates for office must accept campaign contributions from special interests, including members of the development community. For example, in a city of 50,000 residents with a five-member council, campaigning for votes among all residents will be a great deal more expensive than campaigning in a single district among 10,000 residents. In a district campaign, it would be feasible to replace much of the city and regional television and newspaper advertising with personal appearances, bumper stickers and yard signs, candidate forums, and picnics and barbecues, where candidates have to interact with voters in face-to-face situations, answering the tough questions, all at a lower cost than the at-large alternative.

Elections by district have another advantage. In our city of 50,000 residents there are, let's say, 25,000 registered voters. With each of five districts having about 5,000 voters, a group of 100–200 citizens concerned about a development being imposed within their midst in violation of basic planning principles, as in the Old Hyde Park and Southern Showplace cases, would certainly attract the attention (and backing) of *their* council member. That council member, known as "A," could then become an advocate for that group, and would in all probability attract a majority of other council members elected by district, if for no other reason than the fact that these council members might one day need the support of council member "A" when a group of *their* voters approached them in opposition to a NIMBY. Such a scenario would not be possible under at-large representation. The Hyde Park Neighborhood Association could not bring pressure against an *individual* council member because at that time all seven members of the Tampa City Council were elected at-large.[21] Because of that, no member of the council championed their cause. In the Alachua County Southern Showplace case, both the plan amendment and zone change were unanimously passed by the County Commission with no one on this body speaking out in behalf of the local residents. In district systems, I have, over the years, observed that even when *one* member of the legislative body champions a cause favored by local residents, that issue at least gets on the table for debate. The debate usually was won or lost on *technical* merit, rather than sheer power by a special interest waving campaign contributions.

One might ask what about a situation where a NIMBY, such as a jail or sanitary landfill/waste to energy facility, was needed *somewhere* in the jurisdiction? How would this be decided by a council elected entirely by district, where no matter where it was sited, opposition would be championed by at least one council member? I have seen this situation come up over and over across the nation where district representation is in force and the answer is *let the case be decided by planning criteria and related technical considerations!* Then the unlucky council member could look their constituents straight in the eye and say something to the effect of, "Look, I tried, but as you saw in the hearings, the facts dictated that that site was the only one feasible." While voters may be angry for a while, most will be understanding, knowing that the decision, while not to their liking, was at least fair.

We must remember that the reform movements that produced at-large representation and the council-manager form of government were a reaction to the gross corruption of 19th Century political machines, the most notorious being Boss Tweed's Tammany Hall in New York City and Frank "I Am the Law" Hague in Jersey City, New Jersey. Some political scientists have questioned whether these reform movements, championed primarily by the white Anglo-Saxon middle and upper classes, were really for good government or just an attempt to block the rising political power of white ethnic immigrants and later blacks and Hispanics.[22] Regardless of motive, the major issues facing all of us on the dawn of the 21st Century are not centered on the old corruption of graft, patronage, and small favors in local government. Instead, the major issues of today focus on accountability in order to protect against the *new* corruption of poor land use decisions, failed financial institutions, notorious land development schemes,[23] excessive costs for public works due to negotiated or no-bid agreements, and inequities in health care, social service delivery, and education.

Planning Education for Florida's Citizens

The staff report of the FAU-FIU Joint Center on the 1993 Planning and Growth Management Act stated that *"How well the Act succeeds will depend largely on local governments."* This is a profound statement, which implies among other things that only an enlightened citizen can bring to bear enough pressure on local government for these bodies to be serious about planned growth.

In the 20 years that I have watched the evolution of local plans prepared in response to the 1975 and 1985 Acts, the most glaring deficiency was the lack of citizen awareness of planning. Citizen participation was, in 1975, effectively limited to two series of public hearings — one on research and analysis findings, the second on the plan itself. The 1985 Act improved citizen participation, for the record, but the effect was little different in terms of quantity or quality from that in 1975. Virtually no changes in citizen participation requirements were made by the 1993 legislation. The 1975 and 1985 legislation has gone about as far as it possibly can to make citizens aware of the need for planned growth in this state. Planners have long realized that an enlightened citizenry is their best ally in combatting special interests, who would rather not see rational planning take place. Instead of changing the rules for citizen participation, why not have a program of citizen education? A citizenry aware of what planning is about will take the time to serve on advisory boards, attend public hearings, and demand the right to participate by being heard.

In order to get to the next level of awareness — one that will create a demand for quality planning — a citizenry educated as to what planning is all about is required.

History shows us that citizen education is a key factor in planning implementation. One of this nation's most successful planning efforts was Daniel Bunham's Plan for Chicago, adopted in 1909 and implemented by the sale of $234 million worth of municipal bonds during 1911–1931. These bond issues were passed by voters in several municipal elections. Robert L. Wrigley, Jr., tells us that the voters were sensitized to the need for planning implementation by being exposed to the discipline at an early, impressionable age.

Pushing ahead with its promotional work, the Plan Commission published Wacker's Manual of the Plan of Chicago late in 1911. Moody (Walter D.) wrote this celebrated book in an elementary style and illustrated it profusely so that it would carry the message of the Plan of Chicago to 8th grade students. These children were at a very impressionable age and many of them would drop out of school after this grade. In reference to the manual's use in schools, Moody wrote:

> The results of this book appear to be threefold — attracting the attention and sympathy of the parents to the Plan of Chicago...the training of a future citizenry to become responsible in matters of government control; ...the ultimate accomplishment of the whole plan in future years through an enlightened citizenry.[24]

Wrigley went on to state in his article that the Chicago Board of Education formally adopted this text in 1912, and between that time and the early 1920s some 50,000 copies were printed. As to the "Manual's" impact, Wrigley wrote:

> This was the first time a text in planning was admitted for studying by an American school board and it was used for almost a decade with unusual success. Some schools reported that their students begged to have more of this subject and even today, many a Chicagoan looks back with nostalgia to the days when he studied this little book.

Given this background, and present needs here in Florida, I recommend the following:

1. The introduction of planning education into the elementary and secondary social science curriculum throughout the state of Florida and elsewhere.
2. Making a course in planning a statutory requirement for receipt of a bachelor's degree from all of the nine components of a State University System (SUS).

Considerable material now exists to offer planning education at the elementary, middle, and senior high school levels. For the youngest of children, *The Little House* by Virginia Lee Burton[25] will be appealing. It is a tale of a sweet little house that is engulfed by an expanding city. Oppressed by noise, lights, and smoke, the house dreams of "the field of daisies and apple trees dancing in the moonlight."

Several books are available that would be of interest to middle school students. *How Bridges Are Made* by Jeremy Kingston is an account of the design and construction of the five main types of bridges. This work clarifies technical terms and explains facts in a clear, concise manner.[26] *How Maps Are Made* by John Baynes is a history of cartography with easy to read text and illustrations.[27] *Cities 2000* by Robert Royston presents interesting and entertaining hypotheses concerning the future nature and composition of our cities.[28] *Technology 2000* by Peter Evan is illustrated by Isaac Asimov. Evan explains emerging technologies, and speculates on their future applications and societal impact.[29] *How Skyscrapers Are Made* by Duncan Michael, while centering on construction, shows how these buildings relate to their surrounding environment.[30] *How Roads Are Made* by Owen Williams shows considerations that have to be taken into account in the building of roads, including levels of traffic service, topography, safety, and environmental protections.[31]

Other books for younger children include *Little Planner*[32] and *The Treasure of Trash.*[33]

Several books will appeal to senior high school students and young adults. They include *Underground* by David Macaulay, which explores the infrastructure or "root" system of a typical city intersection.[34] *City: A Story of Roman Planning and Construction,* also by David Macaulay, graphically portrays the design of aqueducts in an imaginary Roman town.[35] *Atlas of Environmental Issues* by Nick Middleton focuses on current and emerging environmental issues, including how human activities in the urban industrial and agricultural realm affect the planet's atmosphere, soil and water, and animal and plant life.[36] The best book in this series, in my opinion, is *New Providence: A Changing Cityscape* by Renta Von Tscharner and Ronald Lee Fleming. *New Providence* tracks an imaginary city from early 20th Century prosperity to revitalization in the 1980s. Full color illustrations of cobblestone streets, high-rise buildings, city parks, and residences tell the story of a city in transition.[37]

Computer game enthusiasts will enjoy Sim City, available in Macintosh, PC, and Nintendo formats. Sim City puts the user in charge of a growing city. The player rearranges land uses, zones, balances budgets, installs utilities and controls crime, pollution, and traffic. Just like ordinary citizens "Sims" (the game itself) complains about taxes, traffic, crime, and NIMBYs.[38]

Professional planners can get involved in educating children and young adults about planning. In recent years, the national convention of the American Planning Association has featured a *"planners in schools"* day, where planners and graduate students volunteer to spend time at local elementary and secondary schools teaching planning to these young people. This special day has been held at conventions in Atlanta, Denver, New Orleans, Washington DC, Chicago, San Francisco, and Toronto, between

1989 and 1995, and feedback has been excellent. In Florida, local planners working in cooperation with local school districts could arrange for a similar one-day experience, which would augment planning education by social studies teachers, who would receive training in this area in special workshops conducted by urban and regional planning faculty at the University of Florida, Florida State University, and Florida Atlantic University. The Florida Chapter of the American Planning Association could be a catalyst to connect volunteer planners with public schools.[39]

We must remember that a five-year-old in 1995 will be able to vote in the year 2008, a time at which most of the plans prepared as a result of the 1985 and 1993 legislation will still be well within their implementation period. If planning education by Florida's elementary and secondary students had begun when the Local Government Comprehensive Planning Act was passed in 1975, we would have today a cadre of young citizens exposed to the basics of planning and well prepared to ask penetrating, well-developed questions of planners, appointed officials, and elected officials. The kindergarten kids of 1975 were 25 years old in 1995, and will soon begin to take their places as leaders in our society. We missed a great opportunity in 1975 and 1985; let's not miss another one now.

Why not require completion of a course in planning as a graduation requirement from all units of the Florida State University System? (SUS) and in public university systems in other states? One must consider that nationwide, according to the 1990 U.S. Census figures, less than one adult in seven, or 13.1%, 25 years or older, has completed four or more years of college. For Florida, that figure is 12% according to the 1990 U.S. Census. It stands to reason that adults holding a college degree are more likely than those who don't to rise to leadership positions in their communities. Many will be asked to serve on planning commissions or similar volunteer bodies relating to the planning process, such as economic development boards, community development block grant advisory boards, and other advisory boards dealing with issues such as air and water quality, solid waste, transportation, and social service provision. Just as an accountant's education is broadened by courses in art or music, and the engineer's scope is widened by exposure to the humanities, given the dynamic and complex issues relating to growth management in Florida, elsewhere in our nation, and worldwide, exposure to minimal course work in urban and regional planning is a prerequisite for citizen leadership.

At the University of Florida, such a course already exists. Entitled Preview of Urban and Regional Planning (URP 4000), it is a two-credit upper-division level course that meets twice a week for 50 minutes at each session. The course covers history of planning, standards and criteria for land use, circulation and infrastructure systems, the process of plan preparation and adoption, and an examination of implementation measures such as zoning, subdivision review, capital improvement programs, urban design, and Florida growth management. Offered in a nontechnical format, the course attracts students from a wide variety of disciplines: architectures engineering, business, social science, and fine arts. It is already listed as a general education elective, and making it a required course would not lengthen the number of credits necessary for a degree. Similar courses exist at other state universities in the Sunbelt, including Arizona, California, New Mexico, and Texas.

An investment in basic planning education today will be repaid many times over the years with an enlightened citizenry familiar with the rudiments of planning and growth management. Such a citizenry cannot be easily manipulated by special interests, whether they are diehard no-growth "tree huggers" or single-minded builders who want to pave over everything in sight. Such a citizenry will only be won over by well-reasoned logical arguments, which is the way an educated critically thinking population should respond.

In Closing

These recommendations, based upon Florida's development history, the case studies included here, and this author's personal experiences as a planning practitioner, teacher, and researcher continuously since 1961, are designed to stimulate widespread discussion and even debate. One might argue that only four case studies covering an equal number of governmental jurisdictions is inconclusive; the other 456 units of local government in Florida had no problems with the 1975 and 1985 Acts in producing professionally acceptable plans supported fully by a wide range of citizens.

My experience and intuition say otherwise. When this book is read by a wide variety of planners, elected and appointed officials, citizens and students in Florida and throughout the nation, the majority will say *This happened in my community too!* Then they will debate the recommendations made here and those by others who will soon write on this matter, add some of their own, and we will all be better served by that experience.

Notes

1. "Chiles Declines to Run for Re-election," *Tampa Tribune,* p. A1, March 23, 1988.
2. Among Dr. John M. DeGrove's books are *Land, Growth and Politics,* 1984, and *The New Frontier for Land Policy: Planning and Growth Management in the States,* 1992.
3. Letter of Transmittal, ELMS III final report, *"Building Successful Communities,"* signed by James Harold Thompson, Chairman, and Linda Loomis Shelley, Vice-Chair, December 15, 1992.
4. Recommendations 41 and 42, *"Building Successful Communities,"* Environmental Land Management Study Committee, Final Report, Tallahassee Florida, December 1992, p. 3.
5. Ibid., Recommendations 113–115, p. 4.
6. Ibid., p. 5, par. 3.
7. CS/CS/HB2315, Planning and Growth Management, 1993.
8. Lee James F. Murley, David L. Powell, and Eric Draper. "1993 Legislative Session Update," in *Florida Environmental and Urban Issues,* Vol. XXI, No. 1, Fall 1993, pp. 8–15.
9. Ibid., p. 9.
10. Max Forgey. "Will Growth Management Be the Death of Planning?" in *Florida Planner,* Vol. V, No. 6, July, August 1993, pp. 1, 3.
11. M.J. Matthews and Martin A. Schneider. "The Seventh Annual Growth Management Conference: Intergovernmental Coordination: Still the Weak Link in Growth Management? *Environmental and Urban Issues,* Vol. XXI, No. 3, Spring 1994, pp. 5, 6.
12. Staff Analysis of 1993 Planning and Growth Management, *FAU-FIU Joint Center for Environmental and Urban Problems,* Fall 1993, pp. 14, 15.
13. Ibid., p. 13.
14. Lance DeHaven-Smith. *Environmental Concern in Florida and the Nation,* University of Florida Press (Gainesville), 1991, pp. 29–30.
15. Ibid., pp. 31–33.
16. Ibid., pp. 40–41.
17. Lance DeHaven-Smith. "Overwhelming Support for Land Strengthening Land Use Controls," "The Attitudes of Monroe County Residents on Growth Management Issues in the Florida Keys," *Florida Environmental and Urban Issues,* October 1984, pp. 4–11.

18. Martin Meyerson. "Building the Middle Range Bridge for Comprehensive Planning," *Journal of the American Institute of Planners*, Vol. 22, No. 2, 1956.
19. Frank So. "Strategic Planning: Reinventing the Wheel?" *Planning*, February 1984.
20. Herbert H. Smith. *Planning America's Communities: Paradise Found, Paradise Lost*, Planners Press (Chicago), 1992, pp. 89–110.
21. In 1984, Tampa switched from an exclusive system of election at-large to a mixed system, whereby four council members are elected by district and three at-large. The change came about by pressure from the local chapter of the National Association for the Advancement of Colored People (NAACP), concerned about the lack of black representation.
22. Michael Parenti. *Democracy for the Few.* St. Martin's Press (New York), 1977.
23. The practices of the Florida-based General Development Corporation with regard to speculative land sales was the subject of a "60 Minutes" documentary aired in 1991 entitled *"This House Is For Sale."* Videocassettes of this telecast are available from the Columbia Broadcasting System.
24. Robert L. Wrigley, Jr. "The Plan of Chicago: Its Fiftieth Anniversary," *Journal of the American Institute of Planners*, Vol. 26, No. 1 (February 1960), pp. 31–38.
25. Virginia Lee Burton. *The Little House*, Houghton Mifflin (New York), 1969.
26. Jeremy Kingston. *How Bridges Are Made*, 1985 (Facts on File), p. 32.
27. John Baynes. *How Maps Are Made*, 1987 (Facts on File), p. 32.
28. Robert Royston. *Cities 2000*, 1985 (Facts on File), p. 64.
29. Peter Evan. *Technology 2000*, 1985 (Facts on File), p. 64.
30. Duncan Michael. *How Skyscrapers Are Made*, 1987 (Facts on File). p. 32.
31. Owen Williams. *How Roads Are Made*, 1989 (Facts on File), p. 32.
32. Little Planner (Leisure Learning Products), 1992.
33. Linda Mandel and Heidi M. Mandel, *The Treasure of Trash.* Avery Publishing Group, 1993.
34. David Macaulay. *Underground*, Houghton Mifflin (New York), 1976, p. 112.
35. David Macaulay. *City: A Story of Roman Planning and Construction*, Houghton Mifflin (New York), 1974, p. 112.
36. Nick Middleton. *Atlas of Environmental Issues*, 1989 (Facts on File), p. 64.
37. Renta Von Tscharner and Ronald Lee Fleming. *New Providence: A Changing Cityscape*, Harcourt Brace Jovanovich (New York), 1987, p. 32.
38. All of the books cited in 12-22 and Sim City can be ordered from the American Planning Association's Planners Bookstore at 1313 E. 60th Street, Chicago, Illinois 60637-2891, Telephone 312-955-9100 or Fax 312-955-8312.
39. See the "Teaching about Planning" Section of the 1995 Planners Bookstore Catalogue for several videos and teaching tools about planning oriented toward citizens of all ages.

Index

O

P

R

www.ingramcontent.com/pod-product-compliance
Ingram Content Group UK Ltd.
Pitfield, Milton Keynes, MK11 3LW, UK
UKHW011000180926
475701UK00004B/472
9780367579425